WRECK DIVING TALES

WRECK DIVING TALES

Diving Nova Scotia's Shipwrecks

Allan P. Henneberry

iUniverse, Inc.
New York Lincoln Shanghai

Wreck Diving Tales
Diving Nova Scotia's Shipwrecks

Copyright © 2008 by Atlantis Marine

All rights reserved. No part of this book may be used or reproduced by any means, graphic, electronic, or mechanical, including photocopying, recording, taping or by any information storage retrieval system without the written permission of the publisher except in the case of brief quotations embodied in critical articles and reviews.

iUniverse books may be ordered through booksellers or by contacting:

iUniverse
2021 Pine Lake Road, Suite 100
Lincoln, NE 68512
www.iuniverse.com
1-800-Authors (1-800-288-4677)

Because of the dynamic nature of the Internet, any Web addresses or links contained in this book may have changed since publication and may no longer be valid.

ISBN: 978-0-595-50050-5 (pbk)
ISBN: 978-0-595-61356-4 (ebk)

Printed in the United States of America

The views expressed in this work are solely those of the author and do not necessarily reflect the views of the publisher, and the publisher hereby disclaims any responsibility for them.

Cover Photo: Diver Jason Kennedy swims past the boilers on the WWII Russian freighter Kolkhosnik at 140 feet.

Mike Grebler photo.

*Dedicated to the memory of
Paul L. Crocker, 1956–2006.
A dear friend, a great diver,
an exceptional human being.*

Contents

Author's Note xi
Introduction xiii
Chapter 1:   H.M.H.S. Letitia: A Charmed Life? 1
Chapter 2:   Captain Brannen & the Deliverance 9
Chapter 3:   F.V. Halfish: A Personal Tie 17
Chapter 4:   Forgotten Submarine, HMS L-26 30
Chapter 5:   M.T. British Freedom: Oil on Cold Seas 39
Chapter 6:   S.S. Kolkhosnik: aka The Russian 49
Chapter 7:   Collision at Sea: M.S. Kaaparen 61
Chapter 8:   The "Mystery Schooner"—Hebridean? 73
Chapter 9:   Disaster in Chedabucto Bay: S.S. Arrow 82
Chapter 10: Empire Kingfisher: Risdon Beazley's First Canadian
             Salvage Operation 96
Chapter 11: Halifax Harbour Diving: Bottles & China 113
Chapter 12: Deeper Into the Abyss 127
Afterword 141
Acknowledgements 143
Selected Websites & Reading 145

Author's Note

In the pages of this book you will find references to and photographs of artifacts which were removed from shipwrecks. These items were recovered by divers several years ago when it was commonplace and even acceptable to take home a memento of the dive. It is now illegal to remove artifacts from shipwrecks in Nova Scotian waters except under special circumstances and with the correct permits. Below is information pertaining to the acts which govern artifact removal from shipwrecks in our waters.

The *Special Places Protection Act* of Nova Scotia requires you to have a *Heritage Research Permit* from the Nova Scotia Museum before disturbing any place where historical artifacts are found, including underwater sites. People wanting permits must be able to show a level of skill in archaeology so that important historic sites will be carefully explored and records kept. Such information becomes one source for tomorrow's history books. All artifacts removed from any Nova Scotia archaeological site belong to the Province, that is, to the people of Nova Scotia.

The Canada Shipping Act 2001 (a Federal Act) pertains to all wrecks, including historic wrecks. By this Act, anything recovered from a wreck must be turned over to the Receiver of Wrecks until ownership can be determined. Failure to do so can bring heavy fines. This Act is administered by the Canadian Coast Guard for the Ministry of Fisheries and Oceans.

The provincial *Treasure Trove Act* requires you to have a Treasure Trove License, issued by the Nova Scotia Department of Natural Resources, if you expect to search for "treasure". Even with a Treasure Trove License, you are still bound by the legal requirements of the Special Places Protection Act.

Most local divers will no longer remove anything from the wrecks they dive. Local charter operators will no longer allow divers to remove artifacts from wrecks while conducting dives from their vessels. If, as an individual, you decide to bring back a souvenir of your wreck dive, you do so at your own risk, as you can be prosecuted and heavily fined for your actions. For more information on the subject, consult the above mentioned acts or contact the Nova Scotia Museum.

"Take only pictures, leave only bubbles."

Introduction

Nova Scotia has a seafaring history that is long and varied. From the early Norse seafarers to Sir Henry Sinclair ninety-four years before Columbus sailed from Spain. From the early French explorers and settlers to the founding of Halifax by the British in 1749. From the merchantmen and naval vessels of two World Wars who gave their all (and often their lives) to the fishermen along our coasts who helped feed a growing nation and an expanding world. All these have passed along the shores of our province. All have left a mark: in the landscape, the traditions, and the people. Much of the time, these same seafarers, these adventurers and explorers, have left their mark below the turbulent waters as well, where the bones of their vessels, and often times the bones of the crews themselves, lie in the cold and dark of the ocean's depths until illuminated by a diver's light.

Nova Scotia has almost 7500 km of coastline in total. Beautiful harbours and picturesque bays abound, but there is also a more than healthy dosage of rocks, cliffs, reefs, and small islands thrown in, waiting patiently for the unwary or unlucky navigator. Add to these the dense fogs, snowstorms, and gale force winds which are prevalent along our shores and it is not difficult to imagine why so many ships and sailors have met their end around Nova Scotia's rugged coasts.

All one has to do is drive to Chebucto Head at the entrance to Halifax Harbour on a fine summer's day and stand looking out across the sea. Now imagine the same area during a January storm with darkness, 60-knot winds, blinding snow, and sub-zero temperatures. Imagine a steamer before the age of global positioning, before the advent of radar, trying to make the Port of Halifax in such conditions. Pitching and rolling, the officers unable to accurately fix their position, often carrying seasick passengers or unstable cargoes. It's not difficult then to see why the waters around Nova Scotia's shores are filled with the bones of men and ships that ended their days here. In fact, there are at least eight or ten known and diveable shipwrecks within a stone's throw of the Chebucto Head lighthouse, rusting monuments to the sea's power and man's attempts to bend her to his will.

The causes of these shipwrecks are as varied and unique as the men who sailed the ships. Navigational errors, stranding, adverse weather conditions, collisions with other vessels, enemy activity during wartime, the list of reasons why these vessels foundered is almost as endless as the list of shipwrecks itself.

Nova Scotia has over 8000-recorded shipwrecks, with at least 400 in the region surrounding Halifax Harbour and its approaches. It is suspected that there may be as many as 20,000 wrecks around our shores. Of these, dedicated shipwreck hunters and researchers have found only a small fraction. Many, in the early days of exploration and colonization were never recorded, being simply listed as "lost at sea". The great majority are still lying in wait for some intrepid explorer to find them. A great number of them are broken up beyond recognition. Numerous others, lie at depths too great for divers, even using mixed-gas technology and rebreathers, to reach.

Due to the dedication of a few individuals, more new (to the diving community at least) shipwrecks are being found in our waters each year. Lately, these new wrecks have all been in the technical diving ranges of 150 to 350 foot depths, but each new find shows that the spirit of exploration is alive and well and living in the hearts of many Nova Scotians, as well as others from around the world. Although recent searches have focused on the deeper, more intact World War II wrecks, there are still plenty of wrecks in shallower depths waiting to be discovered by the methodical researcher and patient explorer.

It is however, an uphill battle for these shipwreck hunters. The cost of chartering or buying a boat to be used for a search which could last for several weeks coupled with the expense of specialized search equipment is a factor that is difficult if not impossible to overcome for many would-be wreck hunters. Especially when you realize that most divers and shipwreck hunters are not conducting their searches for financial gain but for the sheer joy of it: the thrill of discovery, the excitement that is felt at being the first to visit a new shipwreck. Added to the (usually) complete lack of compensation is the fact that almost all would be shipwreck hunters are footing the entire bill themselves in pursuit of their passion.

For every new find, there are several shipwreck searches that prove fruitless. There are the countless hours spent staring at a depth sounder's screen, or at the readout of a boat towed side scan sonar. Not to mention the countless hours and days spent in archives and museums researching specific shipwrecks, seeking clues to their locations. The disappointments when searching for shipwrecks are many, the successes few and far between. The rewards however, both in terms of personal satisfaction and historical knowledge, can be immeasurable.

Canada's Atlantic Provinces themselves have a long and noble history of wreck diving that today's divers can rightly be proud of. David Thomas Dobbin of St. Mary's in Newfoundland was probably the first wreck diver on the Atlantic coast of Canada and Newfoundland. He had been a fisherman who worked his own two-masted schooner on the Grand Banks until 1849, when at the age of thirty-two he decided on a career change. Armed with the new and high tech (at the time) Siebe Diving Dress and a hand operated two-manpower air pump to supply him with breathing gas, he would go on to salvage over fifty wrecks in the next fourteen years.

Closer to home was Captain James Farquar in his schooner *Havana* which besides making a living sealing and fishing was also used to salvage the schooner *Alexander R.* in 1906 which had been involved in a collision and sank in Halifax Harbour the year before. However the endeavor ended disastrously when the *Havana*, anchored over the wreck, was herself run down and sank almost alongside the vessel she was salvaging.

And of course the salvage master Horatio Brannen who worked several vessels utilizing hard hat divers in the early part of the 20th century in his vessels the *Coastguard* and the *Deliverance.*

Foundation Maritime also used hard-hat divers in their salvage work from the early 1930's through to the 1970's.

Of course we can't forget the sport divers. Those early intrepid explorers who conducted their dives utilizing Cousteau and Gagnon's new Aqualung. In the 1960's Alex Storm and company found the treasure of the French pay ship *Chameau* near Louisbourg of which much has been written, and another group worked the wreck of the *Auguste* off of Northern Cape Breton for its treasure during the 1970's.

Today, although there is still the odd treasure hunter searching for the wreck that will make them rich, most wreck divers conduct their activities for pleasure rather than profit, but the spirit of exploration and discovery is nonetheless still there. With the advent of better equipment, technical dive training, and the utilization of mixed gas technology and rebreathers, today's explorer's can visit wrecks at depths that those earlier divers could only dream of. But that same earlier spirit of exploration off our rocky coast is alive and well.

This book is not an attempt to improve on that which has been written previously about Nova Scotia's shipwrecks. Nor is it an attempt to chronicle or list all of the known shipwrecks in the area. That would be a lifetime's work and encompass several volumes. This book is simply a compilation of some of my underwater experiences, as well as the experiences of wreck-diving friends and acquaintances coupled with some of the history of the listed shipwrecks.

I have included here the histories of some of my favorite wrecks in the areas of Halifax, St. Margaret's Bay, the Shelburne area, and Cape Breton.

I have also included a chapter on diving Halifax Harbour itself in the pursuit of antique bottles and shipping line china. While not wreck diving, it is connected by the fact that this activity helps keep divers skills honed during those months of the year when wreck diving is an uncertain proposition due to inclement weather. This chapter is by no means a complete tale of diving for artifacts in the harbour and I would direct you to Bob Chaulk's excellent book "Time in a Bottle" to find out more about diving and artifact recovery in Halifax Harbour.

It is my desire to share with you, the reader, some of our local history and some of the remarkable experiences that define Nova Scotia wreck diving. I've always believed, and see no reason to change my opinion, that we have some of the best wreck diving in the world right here in our own watery back yard. I sincerely hope that you enjoy reading about these experiences as much as I've enjoyed reliving them here.

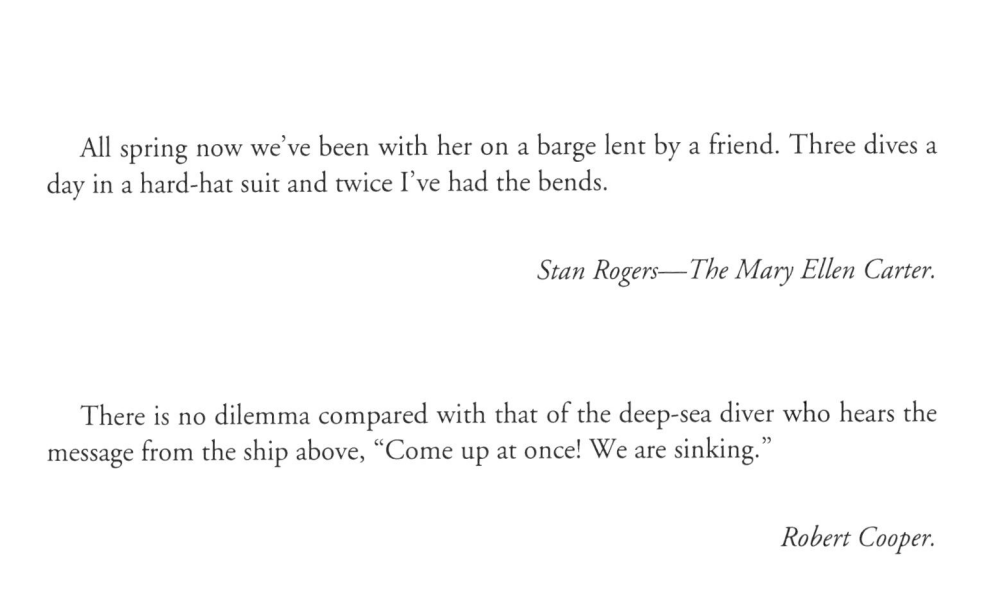

All spring now we've been with her on a barge lent by a friend. Three dives a day in a hard-hat suit and twice I've had the bends.

Stan Rogers—The Mary Ellen Carter.

There is no dilemma compared with that of the deep-sea diver who hears the message from the ship above, "Come up at once! We are sinking."

Robert Cooper.

Chapter 1

H.M.H.S. Letitia: A Charmed Life?

History

The keel of the *S.S. Letitia* first touched the water on February 21st, 1912 on the Clyde River at Greenock, Scotland. Scott's Shipbuilding and Engineering Company, a shipyard with a great and long history going back to it's founding over two-hundred years before, constructed the vessel.

She was built for Donaldson Line Limited of Glasgow, and although not a famous liner like the *Titanic* or *Lusitania,* she was quite a large and comfortable vessel. At 470 feet in length, 57 feet wide, and 8991 gross registered tons, she was well able to cope with the vagaries of the North Atlantic, which was to become her home. *Letitia* was twin-screw, steam driven, and had a top speed of 14 knots. She had accommodations for 300 second class and 950 third class passengers. Along with her sister, the *Athenia,* she was to be put on the Glasgow to Quebec City and Montreal run, mostly carrying immigrants to a new life in the New World. Like other liners of her time, the immigrant trade was to be the *Letitia's* bread and butter.

R.M.S. Letitia *shown here in the liner's traditional black and white paint scheme before her transformation to a hospital ship.*

Paul Grantham photo.

Command of the new ship was given to Captain William McNeill and the *Letitia* departed Glasgow on May 4th, 1912 on her maiden voyage to Canada. But her career as a trans-Atlantic liner was to be short-lived. On August 1st, 1914, Germany declared war on Russia. Two days later, they declared war on France. Britain then declared war on Germany. What was to be called The Great War had begun, and before long the *Letitia* would be called upon to serve her country in the conflict to come.

The *Letitia* was drafted into the fray by the British government almost as soon as hostilities broke out. The liner had no sooner arrived in Glasgow from Montreal after an Atlantic crossing than Captain McNeill was ordered to sail his ship to London where she was to be outfitted for hospital service.

Her transformation included not only the necessary medical supplies and equipment, but also a new paint job. Gone was the trans-Atlantic liner's traditional black hull with white upper works. She was painted entirely white, with a broad green band and red crosses halfway up her hull as was dictated by the Geneva Convention's identification for hospital ships.

With her new look, the grand lady of the Donaldson Line headed for the Straits of Gibraltar and the Mediterranean beyond. She had on board a huge staff of doctors and nurses as well as regular crew to run the ship. She plunged straight into the fray, heading for Suvla Bay and evacuating wounded troops from the Dardanelles campaign. Thousands of Australian, New Zealand, British, and French troops were killed and wounded during the campaign along with uncounted Sikhs and Ghurkas. When all was said and done the Allies had 285,000 casualties out of a total 480,000 men sent to the campaign. Over 60,000 had died, and it was the job of the *Letitia* and others like her to care for the more than 200,000 wounded men that remained. She transported the injured from the front to the Greek islands of

Imbros and Lemnos, and also to Malta, and returned again and again to do her part in the battle against Germany's allies, the Turks.

It was also during this time that she rubbed shoulders with some of the more famous liners of the day. The *Aquitania* and the brand-new *Britannic* had also been requisitioned for war service and were also sent to Gallipoli. These larger ships could not closely approach the beaches to take off the wounded because of their deep draft and this task was often left to the smaller hospital ships such as *Letitia*. They would lie as close as possible to the beach and take on wounded soldiers, who they would then transfer to the larger ships waiting offshore. This would often put *Letitia* and her sisters in direct line of fire.

She seems to have led somewhat of a charmed life during her career as a hospital ship, and was generally considered to be a lucky ship. She traveled at top speed through minefields and watched ships around her die, but she came through unscathed, continuing on her mission of mercy in the midst of the chaos of war.

On one trip from Suvla Bay carrying wounded men, she received a message stating that the *S.S. City of Birmingham* had been torpedoed. McNeill immediately plotted a course to the stricken ship. Giving the new course to the helmsman, she set off at her best speed on another rescue mission. She rescued 350 survivors from the *Birmingham* (as well as the ship's dog) and had barely started back on her course for Malta when another wireless message was received. This one stated that a French ship had been torpedoed and again *Letitia* headed for the danger zone around this new sinking without a thought for her own safety. She again picked up the survivors, over 200 this time. With almost 800 survivors and wounded aboard, as well as her own complement, she finally made her way to Malta. Although German U-boats had watched the ship go about her missions of mercy, they respected the Geneva Convention and made no attempt to harm the hospital ship or interfere with her lifesaving duties.

While working near the front during the assault on the Gallipoli peninsula, *Letitia* was moored as close to the shore as possible without putting her in danger of running aground. With the soldiers in their trenches literally sitting on the beach, there was no room for a field hospital to be set up to care for the wounded, so the *Letitia* had been brought in as a floating field hospital instead.

As the British Navy shelled the Turkish emplacements from offshore and the Turks fired back at them, the *Letitia* sat between the two and went on with her aid of the wounded. The atmosphere on board with those high-explosive shells screaming through the air overhead can only be imagined. But once again, as with her previous life-saving missions, she was unharmed and left to go about her business.

At the end of the Gallipoli campaign, the *Letitia* and several other hospital ships were employed in the evacuation of troops. The withdrawal was completed with surprisingly light casualties, but once again *Letitia* and one of her smaller sisters were directly in the firing line, being used as shuttles to ferry the troops offshore to twelve waiting troopships and larger hospital ships. With the withdrawal from Gallipoli, her part in the war effort was almost complete. The larger Cunard and White Star liners were released from service and sent back to their owners. *Letitia* however, still had one job left to do.

The *Letitia* loaded 456 wounded Canadian soldiers in London in July of 1917. She then headed out into the Atlantic, the ocean she had been built for 5 years previously. Her course was laid for Halifax, bringing the brave boys back to their native soil after surviving the hardships and horrors of the war.

After a mostly uneventful and routine crossing the *Letitia* was approaching the Nova Scotian coast in thick fog on August 1st. Captain McNeill's navigation was correct even with the terrible visibility and the ship arrived at the pilot station off Chebucto Head without incident.

The pilot climbed aboard and made his way to the bridge, where he was given command of the ship to bring her safely to Pier 21 to disembark her passengers. Whether the pilot made an error in judgment in the fog or whether he was actually an inexperienced local fisherman trying to make some extra money (this has been suggested) is not perfectly clear. What is known is that only ten minutes after the pilot boarded and the ship got underway again, the *Letitia* ran aground on the rocks near Portuguese Cove. After coming through all her trials in the Mediterranean unscathed, it seems her luck had finally run out.

Because of the wounded aboard, a long gangway was constructed from the ship to shore. Stretcher cases were carried across the gangway to safety. The rest of the passengers and crew also walked ashore over this bridge. All passengers and crew made it safely ashore with the exception of one crewman who decided to swim to safety instead of taking the bridge and was drowned.

On August 5th, 1917, the *Letitia* broke up, putting an end to the gallant career of this brave ship. At the time of her loss the ship was valued at 1.5 million pounds sterling and would have been a tremendous loss for the owners who were anxiously awaiting her return to their service.

Anchor Donaldson built two more ships with the name *Letitia*. The second one, built in 1925, was also requisitioned for war use, becoming a troopship in

World War II. The third was built in 1961 and was sold to J & J Denholm six years later.

Diving the Letitia

My first trip to the wreck of the *Letitia* was in August of 1993. Two of my dive buddies at the time, Pete DeGrace, Greg MacNeil, and I borrowed an 18-foot Cape Island style boat from a friend and headed out to the wreck, which some thoughtful divers from Dive Masters Scuba Shop had buoyed for us the previous afternoon making the task of finding the wreck easily accomplished.

Jason Kennedy swims past the rudder quadrant and steering gear of the Letitia.

Mike Grebler photo.

After tying into the buoy, we donned our wetsuits, and then struggled into the rest of our gear, all the while sweating in the late-summer heat. Hitting the water felt like diving into a glass of ice water after the heat on the boat, but it was also a relief and we quickly became accustomed to the temperature.

We headed down the line, which was tied off on some wreckage at 50 feet. After checking the pressure in our double tanks we headed east down the slope.

The sheer size and amount of wreckage held us enthralled, neophyte wreck divers that we were. There seemed so much to see, so many places to explore. We swam past cargo-handling winches with cable still wrapped on their drums, huge pieces of hull plating with the holes where portholes had once been showing like empty eye sockets. Our heads swiveled around on our necks as if they were fitted with universal joints.

As we neared 100 feet we came upon the drop off that anyone who's ever conducted a dive on the *Letitia* is familiar with. Although only about 20 feet above the stern of the ship, it can seem like much more due to the darkness often encountered at the site, and doubly so to these three explorers on their first dive to the wreck. We swam out over the drop and let ourselves freefall to the stern of the wreck, adding air to our BCD's to control buoyancy. We landed on the stern deck, complete with bitts, fairleads, and the remnants of the ship's teak decking. We continued swimming out over the stern rail and onto the sand flats beyond, looking back at what remained of the transom of the grand old ship. At 130 feet our air was being used up quickly and our dive computers told us that we were into decompression, so we headed back up the slope, passing the large rudder quadrant and rudder-stock and the tail shaft protruding from the keel, then following the shaft partway up the slope. On a subsequent dive, Greg and I both found items of silver cutlery in this area near the shaft.

We easily found our up line at 50 feet, and ascended to complete our decompression and exit the water. And with that one short glimpse of the wreck of *H.M.H.S. Letitia*, I became captivated with the noble lady of the Donaldson Line.

Over the intervening years, I've returned again and again to this wreck, and the fascination continues to this day even after suffering a case of decompression sickness after a *Letitia* dive and having a friend nearly become a fatality on another. I've conducted hundreds of dives on wrecks that contain more artifacts, are more intact, and seemingly have more to offer, but I always come back and always find something new to hold my interest. Like the once-beautiful teak staircase that I had been told about numerous times but had never seen. Suddenly, or so it seemed, there it was on a part of the wreck that I was sure I knew like the back of my hand. There have been other instances as well, of this wreck surprising me just when I thought I had seen everything on it that there was to see.

Because of the slope on which she lies, the wreck of the *Letitia* offers something for divers of all skill and ability levels. More experienced and/or technical divers will usually head for the deeper portions of the wreck down to 130 feet and swim up the slope at the end of their dives to complete

Jason Kennedy swims over the stern of the wreck, dwarfed by a set of mooring bitts.

Mike Grebler photo.

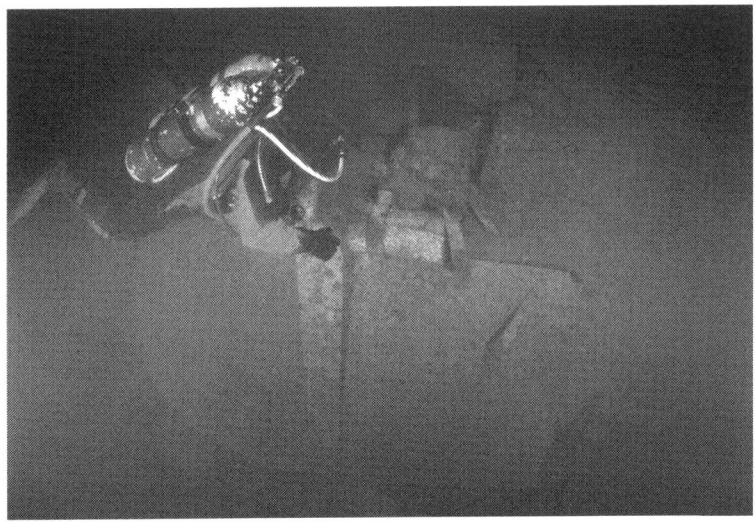

Diver Jason Kennedy inspects the displaced rudder at 120 feet.

Mike Grebler photo.

decompression obligations while still exploring the shallower parts of the wreck. There is some scattered wreckage down to about 140 feet, but nothing really of interest for a few small pieces of rusting hull plate is not usually enough to entice a diver into the extra decompression obligation.

For less experienced divers and those who for one reason or another choose not to dive deep, there are myriad things to see in shallower depths. The wreck actually begins in around 18 feet of water, quite close to the shore and on a fine day with no groundswell doing the shallow part of the wreck can make for a fantastic dive with some amazing photo and video opportunities. Above about the 40 foot mark, the wreck is mostly kelp covered, and for the artifact hunter this can make for a difficult search. At this level there is also a very pleasant swim-through which passes completely underneath the hull where the forward section of the wreck is held up by the rock on which it lies.

Deeper on the wreck, there are still some artifacts found amongst the wreckage. Cutlery, ceramic floor tiles, as well as many brass items are still being found on the *Letitia*. There are still a few portholes to be found, but these are becoming rare. Over the years since my first dive on the wreck, I've recovered two, and have seen two others. It's widely believed among the local wreck-diving fraternity that there are perhaps as many as 200 portholes still waiting to be found, buried under tons of wreckage. These will perhaps come to light as the wreck further deteriorates.

The *Letitia* seems to delight all who dive her. I don't think I've ever heard a diver state after diving the *Letitia* that he or she had seen it all on this wreck and was bored with the dive. Even in reduced visibility, a diver familiar with the wreck can have a great dive. Paul Crocker and I once did a dive to the wreck in visibility so bad that it was like night at only 50 or 60 feet of depth. We elected to continue our dive and had no problem navigating the wreck because we were so familiar with it. Although most of the divers on the boat that day were complaining of the visibility and that they'd had to cut their dives short, we both stated that even though the visibility was so bad, we had had a fantastic dive. The *Letitia* is that type of dive, and it'll keep you coming back time and time again.

Chapter 2

Captain Brannen & the Deliverance

History

Horatio Harris Brannen was born on December 27th, 1872 at Wood's Harbour, on Nova Scotia's south shore, the son of William and Virginia (Nickerson) Brannen.

Before he was 20 years of age, he was sailing as master of deep-sea fishing vessels out of Clark's Harbour on Cape Sable Island. His exceptional navigational expertise soon led to Brannen sitting the examination for his captain's and eventually Master Mariners tickets.

Around the turn of the century, this enterprising young man caught the eye of the Barrington Wrecking Company, who offered him the job of captain on their salvage vessel *S.S. Coastguard*. With his salvage gang and divers, Captain Brannen learned his new trade well and became a very successful salvor, being richly rewarded working the wreck of the *S.S. Hungarian,* a mail steamer that was lost near Cape Sable in 1860 along with salvaging many other wrecks. Also among his many exploits was the refloating of *H.M.C.S. Niobe* when she went on the rocks of Cape Sable Island in 1911.

When the Southern Salvage Company of Liverpool, N.S. took over the operations and assets of the Barrington Wrecking Company, including the *Coastguard*, Brannen went along as well. Southern Salvage later gave him command of the

S.S. Deliverance, a larger and newer vessel, and he remained her skipper until she sank from under him in June 1917.

Captain Brannen and the *Deliverance* were both taken into the service of the fledgling Royal Canadian Navy at the outbreak of World War I, and Brannen became as proficient at his new job in naval operations as he had been at his other endeavors. Although engaged as a naval auxiliary vessel, it seems that Brannen and his ship were still involved in commercial salvage and diving work when the opportunity arose.

After the sinking of the *Deliverance* Brannen was given command of the 125', 228-ton steam tug *Stella Maris*. As with the *Deliverance* (and Brannen himself), she also was under charter to the navy, her peacetime owners being the Halifax Graving Dock Company. The *Maris* had originally been built as a gunboat for the Royal Navy in 1882. After being decommissioned from the Royal Navy in 1911, she was converted to a salvage ship by a St. John's company and was eventually purchased by the Halifax Graving Dock Company in 1915.

One of Brannen's first jobs as her skipper was to steam to Newfoundland to rescue the stricken Norwegian America Lines vessel *S.S. Khristianafjord* of 10,670 tons (a huge ship for those days, and ultimately a huge salvage award for Southern Salvage and the *Stella Maris*' crew) on the 14th of July 1917.

Not quite five months later on the morning of December 6th, 1917, Captain Brannen was towing two scows from Pier 8 in Halifax into Bedford Basin when the collision between the Belgian relief ship *Imo* and the French ammunition vessel *Mont Blanc* occurred near the Narrows. Hurriedly casting off and anchoring his two charges, he immediately headed for the flaming *Mont Blanc*. His plan, with the help of several naval hands who had arrived on the scene, was to get a towline on the burning vessel and haul her away from Pier 6 where she had drifted, burning after the collision, while at the same time ordering his crew to turn the tug's fire hoses on the flaming ship. The fire fighting equipment that the *Stella Maris* carried was completely inadequate to deal with the floating nightmare of explosives that was the *Mont Blanc* (she was fully loaded at the time with a cargo of benzene, picric acid, gun cotton, and TNT), and the towline was still being prepared when she blew. Halifax was devastated by the World's largest pre-atomic explosion, with most of the city's north end flattened and over 2000 people killed. Captain Brannen and 18 of the *Stella Maris*' crew of 24 were also killed in the blast. One of the 5 survivors from the *Maris* was Brannen's son Walter, who had been sailing on the tug as his father's first mate.

The *Stella Maris* herself ended up aground near where Pier 6 had been before the explosion occurred. She was later refloated, repaired, and remained in service for the next ten years.

S.S. Deliverance *shown here shortly before her demise.*

Paul Grantham collection.

The *S.S. Deliverance,* official number 131204, was built by the Southern Salvage Company at their shipyard in Liverpool, N.S. and launched in 1914. She was to be a new command for Captain Horatio Brannen, replacing the old *S.S. Coastguard.* She was 111 feet in length, with a beam of 32 feet and a draft of 10.5 feet. She displaced 280 tons. Something rare in the small coastal vessels of the day, she was a twin-screw vessel. Her two bronze propellers were driven by two triple-expansion steam engines, which in turn were fed by a huge single, high-pressure boiler.

The *Deliverance* was built of Nova Scotian native woods, with a hull of softwood over hardwood timbers in much the same style as the local fishing schooners, but there any similarity to the beautiful and graceful fishing vessels ended. Perched atop her main deck aft was a blocky, two deck superstructure comprising her wheelhouse, officers quarters, galley, and engine and boiler room casings. From the top of this superstructure protruded her single, ramrod straight funnel and engine room ventilators. She had a large, open forward working deck for conducting salvage operations and a small deck area aft of the superstructure, set up to deploy and retrieve her salvage divers when working wrecks.

The vessel was to be used for diving and salvage work but she was requisitioned for war use soon after the outbreak of World War I and became a naval minesweeper and auxiliary vessel, put to use as a Jack-of-all-trades at whatever tasks were required. It was in this role that the three-year-old vessel found herself on the morning of June 15th, 1917. The *Deliverance* was steaming out of Halifax Harbour when she was struck and holed by the inbound Norwegian barque *Regin* near Portuguese Cove.

With water pouring into the wound created by the bow of the *Regin*, Brannen realized his ship was doomed if he couldn't get her into a sheltered harbour where she could be beached before the water reached and quenched the fires for her boiler. The coastline around the Portuguese Cove area being made up of granite boulders and cliffs, Brannen made his decision. The nearest place where the *Deliverance* could safely be beached was the tiny fishing village of Herring Cove.

Ordering full speed on both engines he brought his stricken vessel around and set course for Herring Cove. With the twin engines turning out their best speed, and the pumps already losing their battle with the incoming water, the *Deliverance* drove on towards the Cove.

It was a brave and valiant effort on the part of Captain Brannen and his crew. But it was easy to see that she was already doomed. Just outside the mouth of Herring Cove, near where the navigation buoy for entering Herring Cove sits today, the *Deliverance* gave up the ghost and slid beneath the waves. The water had reached and doused the fires for her boiler sometime before, so there was no longer any steam to run the engines or pumps. With the death of those flames, so also died the hope of saving the little vessel. The crew abandoned the *Deliverance* at Brannen's command. Her captain however, unable to believe that he couldn't save her, stayed aboard and fought for her life until she slipped beneath the surface. As if unable to let go of the man who had been at her helm since she was launched, the *Deliverance* very nearly dragged him under with her when she went. Only three years after she had been launched, the *S.S. Deliverance*, diving tender, salvage ship, and H.M.C. auxiliary vessel met her end in a watery grave at the mouth of Halifax Harbour.

Diving the S.S. Deliverance

In her death, the little salvage ship has bestowed the gift of life. For she rests in a desert-like wasteland of fine, hard packed silt where almost nothing grows. The wreck itself however, like a desert oasis, is teeming with life. Green sea urchins, sea cucumbers, and starfish abound. The remains of her hull planking are covered in anemones. Wolf eels, crabs, lobsters, perch, redfish, cod, and pollack often call the *Deliverance* their home.

The wreck of the *Deliverance* sits upright on the bottom. It lies beneath 115 feet of cold, dark North Atlantic water. It always seems, for some inexplicable reason, that the water temperature is much colder on this dive than on other wrecks in the area. Even in September, when water temperatures are the warmest of the year, it's cold on the *Deliverance*. There is a freshwater stream emptying into Herring Cove from inland. While this may have an effect on temperature, I've always felt that the wreck was too far away for this to make any significant difference. Whatever the reason for this seeming temperature drop, bring your argon bottle for drysuit inflation on this dive; you'll be glad you did.

Fine sediment, often suspended in the water column, usually makes the *Deliverance* a dark dive with poor visibility. Mike Grebler, our dive group's resident photographer, spent years diving the wreck and trying to obtain good photographs of it before he finally hit on a day with terrific visibility in the summer of 2006. Only twice while diving this wreck have I had visibility in the 30 plus-foot range. More common is for visibility to be in the range of 5–10 feet. There have been a few dives when I've touched down on the wreck without seeing it at all.

On those days when you can see it however, it is certainly an enjoyable dive. The hull itself is mostly intact, although for the most part buried in the mud. The decking and superstructure have long since disappeared with the exception of one small portion of her afterdeck. Collapsed across this piece of decking aft of the engine room is one of the gallows frames used for heavy lifting when conducting salvage work and also for towing her minesweeping gear. Not far away, at the very stern of the ship lies a hardhat diver's umbilical and lifeline, now covered in coralline algae as if encased in cement, but still neatly coiled and ready for use just as some long ago diver's tender had left it. Over the years, there have been odds and ends of the *Deliverance's* dive equipment recovered from the wreck. A visiting diver from New Jersey found a hard-hat diver's knife in the mid-1980's. A retired diver of my acquaintance once recovered several dive helmet faceplates. Another friend unearthed a standard diver's breastplate, which unfortunately was lost while he was completing his decompression at the end of the dive.

Above: *The anchor windlass on the wreck of the* Deliverance. **Below:** *The author examines the encrusted divers' umbilical near the stern of the wreck.*

Mike Grebler photos.

Swimming forward from the afterdeck will bring you to the area of the ship's engine-room with it's twin triple-expansion steam engines. The lower ends of the engines and propeller shafts are buried in the silt, only the upper 4–5 feet of the engines are visible. Continuing forward, you will come to the single boiler, it's large size making it seem to be from a much larger ship than the *Deliverance*. The boiler's diameter is nearly as large as the ship's width and must have allowed the engineers only the narrowest of passages around it when traveling forward or aft through the boiler-room.

Heading forward beyond the boiler is a flat, open expanse of sediment corralled by the sides of the hull. This is the area of the *Deliverance's* working deck. On those rare days of good visibility, you can see the sides of the hull to either side of you as you swim towards the bow. Unfortunately, much of the wreck is buried in the sandy bottom at this point.

Lying at a right angle to the wreck at this point off the port side is the ship's foremast. It was while swimming along this feature in 1986 that an American diver discovered the bell of the *Deliverance*, still attached to the mast by its bronze fittings. The diver subsequently recovered the bell and took it with him home to New Jersey.

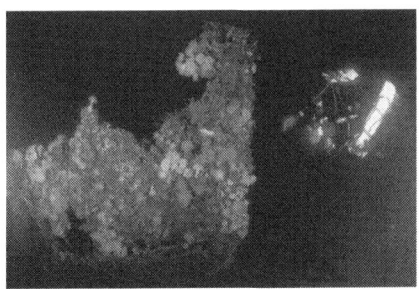

The author hovers near the bow of the Deliverance. *The sterile sand bottom can just be made out beneath the diver, in stark contrast to the profusion of anemones adorning the wreck's hull planking.*

Mike Grebler photo.

Reaching the bow, you will find her anchor windlass sitting in place, waiting for the bosun's hand on the steam valve to start the anchor chain clanking inboard. Her stem stands upright at the triangle of her bows, as if still thrusting on towards Herring Cove and the safety of a dry beach on which to lay herself.

With the *Deliverance* being only 111 feet in length, it doesn't take long to see the entire wreck. You can take your time swimming from the stern to the bow and back again and still remain within your computer's no-deco time. For the technical diver, a longer dive (I have sometimes spent an hour or more on the wreck, poking around in the engine-room area) can be planned, perhaps brushing away sediment to expose an artifact or two, or swimming further off the wreck looking for parts of the hull, superstructure, or rigging that may have been jarred loose by or collapsed after the sinking.

There's something about this wreck that has always intrigued me. Perhaps it's that umbilical coiled so neatly on the stern, knowing that working divers once trod her decks. Or maybe it's the thought of her captain trying so valiantly to save her, and almost losing his own life in the process; surely she meant something special to him? Whatever the reason, every so often I hear the call of that bell buoy near Herring Cove, and will spend several days diving this small wooden shipwreck at the mouth of Halifax Harbour.

Although the *Deliverance* lies within sport diving limits, it is not a dive to be undertaken lightly. The poor visibility and darkness can be disorienting to divers with a low tolerance to narcosis. The no-deco time is short, there are often strong currents, and the wreck is polluted with old ropes and fishing nets. And there is the cold, making a dry suit a must for all but the hardiest (craziest?) divers.

In spite of all this, and because of some of it, the wreck of the *Deliverance* is an eerie, intriguing, and very beautiful dive.

Chapter 3

F.V. Halfish: A Personal Tie

History

As a small child growing up in a fishing family in Eastern Passage at the mouth of Halifax Harbour I often had occasion to play near the ocean, on the tidal flats and in the tide pools near my father's wharf. At times I was allowed to play in one or other of my father's boats, fancying myself an intrepid mariner, braving sea and storm, reef and rock to bring my catch safely to port. Such is the imagination of a child.

Of course I was also told tales of the sea, of fishing, and of Devil's Island (the Henneberry's ancestral home since leaving Ireland in the early 1800's, it's a tiny island lying at the mouth of Halifax Harbour) by both my grandfather Edmund and my father Edmund Jr.

One of the stories told to me by my father was of the time he and my grandfather had found two dories adrift while out tending their mackerel nets one day in the fall of 1966. The tale goes something like this:

"Dad and I were out to the south'ard of The Island (Devil's Island) *hauling mackerel nets, trying to catch enough to salt down for bait for the upcoming lobster season. Dad saw something off in the distance near the Navy Buoy* (Navy West, a buoy used to mark the western end of the target tow corridor for naval practice firing) *and asked what I thought it was. My eyes being better than his, I could make out what looked like a half-sunken dory and said as much. We decided to take a closer look and upon arriving*

alongside found not one, but two double-dories (sixteen foot bottom length, usually operated by two crewmen) *nested together* (for those not familiar with these amazing little sea-boats, the seats and other fittings could be removed from them so they could be stacked inside each other thus saving precious deck space when stowed on board ship). *They had come from a side dragger which caught fire and sank near Chebucto Head the previous day. We took them in tow for home where we soon sold one to a neighboring fisherman and kept the second one for ourselves."*

That second dory was one of the boats in which I whiled away hours as a child, dreaming of open seas and far off lands. It remained in our family until 1980 when it was sold to a friend of my eldest brother. Little did I realize as I played at being the salty old sea captain that the dory and my father's tale of how he came by it would factor into my diving activities thirty-some years later.

The fishing vessel *Halfish* was a Fairbanks-Morse diesel powered side dragger that had been built in the little Nova Scotian port of Mahone Bay in 1944. She was 100' overall with a beam of 24' and displaced 90 tons. She had a steel superstructure nestled atop a wooden hull of pine planking over oak timbers. Her

The Halfish *in the summer of 1965. The dragger is shown here aground on a sandbar in Yarmouth Harbour. The four dories in two nests of two can be seen aft of the wheelhouse.*

Photo courtesy of Hubert Hall, Shipsearch Marine.

superstructure was located right aft with the wheelhouse perched atop its forward end in the style of the draggers of the time. The vessel had been built for Halifax

Fisheries Ltd. to be engaged in the lucrative fresh-fish trade, at that time taking over the reins from the old saltbanker schooners who spent months on the fishing grounds and wet-salted their catch at sea. She towed an otter trawl to catch groundfish, its mouth kept open by two 1500 pound doors, the warp wires for which were hauled in by huge split-winches forward of the superstructure on the main working deck. The mainstay of her catch for the next twenty-two years would be codfish, brought ashore fresh and packed in cracked ice, to be filleted and sold fresh or frozen or split and salted for the foreign markets.

On October 14th, 1966 at around 4:45 pm, fire broke out aboard the aging *F.V. Halfish* while twenty miles off Lunenburg, Nova Scotia. At the time, the dragger had been heading from Lunenburg to Halifax to pick up the rest of her crew and would then have been leaving for a fishing trip to Western Bank, about ninety miles offshore. Shortly after leaving Lunenburg, the main engine broke down, an occurrence which had happened frequently in the past and had led to the vessel being towed in from the fishing grounds three times previously since June.

Captain Bernard Hebditch ordered chief engineer Hubert Conrad to start the auxiliary generator to give power to the radio telephone so he could place a ship-to-shore call to the vessel's owners alerting them of the situation. This was done, and the chief returned to the wheelhouse to tell the captain he now had power to make his call. Returning to the engine room doorway only a few minutes later, Conrad found smoke billowing from the opening and sparks shooting from overhead wiring. He gave the alarm and all the fire extinguishers the *Halfish* possessed were brought to bear on the fire. Meanwhile the captain set off distress flares.

Nearby was the Canadian Navy auxiliary tug *Riverton,* which came in response to the flares, tied alongside and helped to fight the fire. All of the *Riverton's* fire extinguishers but one were used to fight the flames. By 5:30, the job was done. *Riverton* evacuated six of the *Halfish's* crew, leaving aboard only Captain Hebditch and his first mate.

That night at around 11:00 pm fire again broke out on the vessel. The destroyer *HMCS Kootenay* had been standing by since *Riverton* had begun the tow and she now put a firefighting crew aboard the *Halfish*. After a battle lasting into the wee hours of the morning, the persistent flames were once more extinguished. But as all mariners know, a fire at sea, that most feared of all shipboard hazards, is insidious. As the little flotilla arrived off Chebucto Head and prepared to enter Halifax Harbour early on the morning of October 15th, fire erupted on the *Halfish* for a third time. The *Riverton* and *Kootenay* were now joined by the naval auxiliary tug *Glendyne* which had just steamed out from the Halifax Dockyard to take over the tow from *Riverton*. The naval fire tug *YTM 556* was also called in and came alongside the stricken dragger to fight the fire with her water cannon and fire hoses but it was too late. There had been too much damage done to the little fishing ship by the fires, and too much water pumped aboard her to fight them. At 10:18 am on Saturday, October 15th, 1966 the doomed *Halfish* slipped stern first beneath the waves two miles east of Chebucto Head.

There are two naval photographs which were taken from the *Kootenay* which were released in the Chronicle Herald on October 16th. The first shows the *Halfish* on fire and engulfed by smoke while the *YTM 556* pours a stream of water into the dragger from her water cannon and the *Riverton* stands by. The second photo shows the sinking *Halfish* with just a few feet of her bow left above water as she goes down by the stern.

The crew of the *Halfish* were all landed safely in Halifax by the *Riverton* and then made their way overland to their homes on the province's south shore. Unfortunately for some, this incident would not be the last brush they would have with disaster at sea. Just four short months later, on February 21st, 1967 (reported as Black Tuesday in the Chronicle Herald and Mail Star) there were three Nova Scotian fishing vessels sunk with great loss of life in a terrible winter storm. On that day, the National Sea Products dragger *Cape Bonnie* was wrecked on Woody Island with the loss of all hands. Captain Bernard Hebditch, a native of the French islands of St. Pierre et Miquelon, was one of those lost, having signed on the *Cape Bonnie* as mate after the loss of the *Halfish*. He was only twenty-six years of age.

At the time of her sinking, the *Halfish* was owned by Wedgeport Canners Ltd. They had purchased the vessel from Halifax Fisheries Ltd. the previous year.

In the summer of 1965, the *Halfish* had run aground in Yarmouth Harbour, only to have this ignominious moment captured on film by Hubert Hall. So it seems that although the *Halfish* had performed admirably for those who originally owned her, she turned out to be an unlucky ship for her final masters.

I spoke to the owner of Wedgeport Canners, Mr. Alfred LeBlanc, by phone. At the time of this writing he still lives in Wedgeport and now owns and manages Chebogue Fisheries in Yarmouth. While he couldn't tell me much about the vessel, she being only one of a fleet and with the company only a short time, not to mention that the sinking happened forty years ago, it was he who gave me the name of the captain of the *Halfish* as well as relating to me the tale of his tragic demise on the *Cape Bonnie*. I did find it a pleasant change as a shipwreck researcher however, to be able to speak with someone still alive who had been involved in the incident that I was investigating, as most of the wrecks that I and others research happened too far in the past for those who were involved to still be around.

Diving the F.V. Halfish

In the summer of 2005, my good friend Jason Kennedy, while surfing the internet looking for bits of information on obscure or little reported shipwrecks, happened upon a reference to an ROV (remotely operated vehicle) survey conducted by the Canadian Navy in 2004 as a test run for a new ROV they were in the process of acquiring. Having been given the position of an anomaly near Chebucto Head by Gordon Fader, who had bathymetrically mapped the area while working with the Geological Survey of Canada, the team found there was indeed the wreck of a small ship at the position given. The ROV recovered the vessel's propeller, which oddly enough allowed the team to positively identify the wreck, as it had her name—*Halfish*-stamped into the bronze of the hub.

Jason promptly sent off an email to Glen Canning, who had been the ROV operator during the 2004 test, and subsequently obtained the position of the wreck from him. Never before or since, even after a lifetime of commercial fishing, have I seen a more exact position: the GPS latitude and longitude were given to four decimal places. This was just too good to be true!

Although none of the divers in our immediate group had heard of any vessel called the *Halfish,* a little digging turned up the information that she had been a side dragger that had caught fire and later sunk near Chebucto Head at the mouth of Halifax Harbour in 1966. It was one of those things that we talked about on the dive boat when we were heading out to dive other wrecks. For over a year we talked of it on occasion, saying things like," We should dive the new wreck that Jason got the position for." But other wrecks and projects beckoned and so the *Halfish* would for the time being remain a virgin wreck even though we had a confirmed position for it.

Finally, on November 24th, 2006, the time for our procrastination was over and the little ship's time had come. Four divers and two crew set out on the dive vessel *Lady Shirleen* to the reported position of the *F.V. Halfish*. The day was overcast but the seas were moderate with the strong westerly wind coming off the land. Nova Scotia wreck diving demands strict observance of weather conditions but we had gotten lucky with an almost perfect day this late in the season. Although all those aboard were seasoned technical wreck divers, only four of us would be diving on this day. Two of our group, Jeff Smith and Cameron Fraser, had elected not to dive today and instead would act as boat crew and surface support for the rest of us. All four divers had prepared their gear on the hour-long steam to the site and were anxious and ready to dive: all we needed now was for a shipwreck to appear at the position we had been given.

As we approached the reported position all hands crowded into the *Lady Shirleen's* wheelhouse to watch the depth sounder as it traced the ocean bottom over which we passed. We closely watched the bottom profile as we neared the position. At the exact coordinates given to Jason by Mr. Canning the appearance of a wreck on the bottom began to be traced across the screen. Divers and crew looked expectantly at each other but in the back of everyone's mind there remained a nagging doubt. Was it actually a wreck? We had all been misled in the past by soundings that just had to be a wreck only to find upon diving that we had just been the first divers to visit a pile of boulders. Would this one be different or turn out to be another disappointment?

Running back across the position, Jeff yelled for Cam to drop the anchor and stopped the boat at the appropriate spot. We waited for the wind to tighten the anchor line to see if it would hold. As the line came tight, Cam and I each placed a hand on it where it passed over the rail. Then I quietly spoke, "It's dragging." The groan from the assembled divers was palpable. Maybe this was just another pile of rocks. Perhaps the position given to Jason by Glen Canning was erroneous after all.

Jeff said he'd motor slowly over the position again, this time dragging the anchor and hopefully catching the wreckage. We again crossed the position, saw what appeared to be wreckage, and Cam and I again held the anchor line as Jeff took the dive boat out of gear and let her forward momentum slowly drag the anchor across the bottom. Then: the anchor caught, dragged, caught again and held. We had hooked into something that held an anchor better than a pile of boulders would.

Having discussed it on the way to the site, it had been agreed that Jason and I would splash first to do the tie-in if in fact there was a wreck here. We quickly geared up with help from our intrepid crew. Sam Millett and Nick Seemel would follow us ten minutes after we left the surface. If we weren't back on the surface within ten minutes it would signify that there was indeed something here worth taking a look at. I jumped from the stern of the boat first, quickly followed by Jason, and we began our descent along the anchor line to whatever was waiting for us below.

At a depth of just over 100 feet, I was beginning to make out what appeared to be man-made shapes in the gloom 50 feet below me. I turned around and signaled to Jason above me with my light. Giving him two thumbs up I pointed my light downward to where I could now make out the distinct outline of a small ship.

Arriving on the foredeck of the upright wreck I swam to where the anchor had snagged a piece of wreckage in order to check that it was secure and wouldn't pull out during our dive. I glanced at my bottom timer: 149 feet. While I moved the anchor to a better holding spot, Jason grabbed the two fathom tail rope and tied the anchor in to an exposed but still solid oaken rib of the vessel. Now that the anchor was properly secured, we could begin our exploration.

What we discovered was a wooden hulled vessel with a steel superstructure aft and the remains of a large working deck forward. With visibility in the 40 foot range we could take in much of the vessel at a glance. The vessel's configuration as a side dragger was evident and her fishing gear was still aboard, although torn and tattered after forty years in the sea. The huge net was draped across the foredeck where it had been stowed at the beginning of her final trip, and the large trawl doors (or otter boards) which keep the mouth of the net open were still in their places near the gallows frames on her starboard side.

We swam aft along the port side of the steel superstructure, noting points of interest and hazards to be aware of for future explorations. The port side fuel tank lay in the sand beside the wreck's hull adjacent to the forward end of the engine room. We arrived at the stern and as we rounded it we could see the rudder still attached to it's stock but wrenched to the side and partially buried in the sand. The prop was of course conspicuously absent, having been removed by the ROV in 2004. We then swam along the starboard side towards the bow, noting that most of the forward section had collapsed, leaving the ribs along the vessel's side exposed and sticking up above the foredeck. We passed the split open bow and stem and again turned to swim aft, this time over the midships line of the wreck, with Jason making a quick side trip into the engine room entrance while I continued further aft and did a short swim-through of the stern section.

The hull of the wreck was badly deteriorated from the combined effects of the harsh North Atlantic currents and weather and the tireless hunger of wood boring marine organisms. There was enough relief aft however, to be able to look in (and swim into) the engine room, noting the large Fairbanks-Morse diesel engine with the main and faulty auxiliary generators mounted one on either side of it.

By this time Sam and Nick had arrived on the wreck to begin their dive. As they began their exploration of the wreck, our bottom time was nearing its end. After 25 minutes at 150 feet, it was time for us to head for daylight. We swam forward again, towards the anchor line, passing over the large split winches used to haul in the hundreds of feet of wire warp attached to the net when fishing.

We ascended to our first deep stop at 95 feet where we were buzzed by a playful and curious seal who was no doubt wondering what these bubble-blowing intruders to his realm were and had stopped by for a closer look at these strange creatures. Having satisfied himself that we posed no threat and deciding that we

Above: *One of the many brass fire extinguishers which litter the wreck of the* Halfish. Author photo. **Below:** *The main fish hold hatch on the working deck with the access ladder sticking up over the hatch coaming. The deterioration of the wreck can be plainly seen in the deck planking.*

Paul Grantham photo.

were probably inedible, he soon swam off into the gloom in search of a sashimi lunch or perhaps some more entertaining playmates.

We continued our ascent, completed our chilly decompression uneventfully in the 34 degree Fahrenheit water and surfaced 55 minutes after we'd submerged wearing ear to ear grins on our faces. It wasn't a pile of boulders after all. We had just completed a dive on a virgin wreck, perhaps just an old side dragger, but still a first for both of us.

Being part of the first dive team on a virgin wreck is to a wreck diver a thrill like no other. But the wonder of this wreck for me had come the day after the dive. As I'm sure you've deduced, the two dories found by my father and grandfather in 1966 were two of the four that the *Halfish* had carried as her lifeboats. These had floated off and drifted away after the dragger sank. Over the years I had forgotten this tale told to me by my father when I was a child. It was only when I was relating the tale of our dive to my eldest brother the next day that realization came to me. As I told him some of the history of the *Halfish* he remarked, "That's the dragger that Dad and Granddad picked up the dories from."

Forty years after my father and grandfather had picked up those two dories from the stricken dragger, I of all people, turned out to be the first sport diver to visit the wreck of the *Halfish,* thus coincidentally tying three generations of my family together with this unfortunate Nova Scotian fishing vessel.

On our next dive to the wreck of the *Halfish* on January 7th, 2007, Jason and I again entered the water first to set the hook. As I neared the bottom, I could make out large amounts of netting billowing above the wreck. This somehow looked different from our last dive; the netting was much higher in the water column. I soon found the reason why: we landed on a separate portion of the wreck away from the main body, comprising the area of her foredeck where the mast and fore gallows were located. The net was entangled in the still upright gallows and it was these that were holding it up from the bottom. This displaced area of foredeck was about 15–20 feet in length and the width of the vessel's deck. For it to be wrenched completely off the ship and deposited northwest of the wreck is testament to the power of the storms in the area, that even at this depth (and greater, as we found in 2003 on the wreck of the *British Freedom)* the storm surge could exert such force.

I quickly tied in the anchor to an exposed steel ring bolt and we set off in search of the main body of the wreck. Swimming to the southeast, we quickly spied another small piece of foredeck, with two round deck hatches attached to it, used for putting the catch through to the fish hold without the need to open the main hatch. Next came the starboard side fuel tank, like the port one lying on its side in the sand. By the time we passed this large steel box, the main body of the wreck came into view. We arrived on the forward part of the wreck on the starboard side about 150 feet from the displaced portion of foredeck where we'd tied in the anchor.

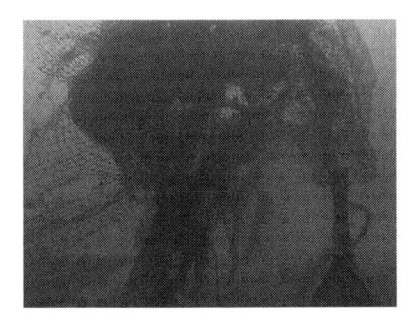

Left: *The port side of the wreck alongside the deckhouse. The deck has deteriorated, showing the machinery that was once hidden beneath it. In the background can be discerned the after port-side gallows, the heavy steel frame from which the net was towed.* **Right:** *Part of the net of the* Halfish, *billowing up from the displaced portion of the wreck's foredeck.*

Jason Kennedy photos.

One thing I noticed on this dive that I had missed previously was the profusion of brass fire extinguishers. These lay scattered in odd places among the wreckage and it was obvious that there were many more than a vessel this size would carry. We were seeing the *Riverton's* efforts here in her discarded fire-fighting apparatus.

Continuing aft, we crossed to the port side, my objective on this dive being to photograph the wreck. With this in mind I was heading for the engine room doorway, thinking to capture on film an image of the faulty gen set. On this dive it was not to be: Jason, who had been planning to document the wreck on video, had problems with his camera while still on the boat and there it remained. Now it was my turn and my digital camera seemed to also have a ghost in its machine. The camera wouldn't remain powered up for more than a few seconds, not nearly long enough for me to get a shot. Time and time again during my dive I tried to get the stubborn imager to fire, but it had made up its mind and wasn't going to be cajoled into doing its job. Oh well, Jeff Smith would soon arrive on the wreck with his trusty Nikonos V film camera and could get the shots I needed.

I was heading back to the upline to begin my decompression and help point the next team towards the main body of the wreck when I met Dwayne McLaughlin descending to his first encounter with the *Halfish*. I showed him where the wreck lay, and then headed for the surface. When I had completed my first gas switch at 90 feet, I again tried my camera: powered up, pointed it at Jason to take a shot, and again it shut itself down. Damn thing! Where was Jeff by the way?

Again at 30 feet, I made my gas switch, and tried the camera. This time it worked. Great, the dive's over and now the camera's working. I took a few photos of Jason and later Dwayne as they completed their decompression. Small consolation.

Jason Kennedy videos the bow section of the wreck, swimming past oak ribs exposed by the foredeck's collapse and displacement.

Author photo.

I surfaced after completing my deco and was met at the *Lady Shirleen's* stern ladder by Jeff. It being a bitter cold January morning and he being chilled before we even arrived at the wreck site, he made the wise and rational decision not to dive. There sat his camera on the dive boat's captain's chair, unused like the others. Photographing this wreck, far from being the usual simple procedure, was proving to be a trying and troublesome task. Oh well, as we often say when wreck diving, "The wreck's not going anywhere, there's always the next dive."

Jason Kennedy adjusts his buoyancy at the twenty-foot deco stop after the second Halfish *dive.*

Author photo.

Finally, on May 21st, 2007 we began to obtain images of the wreck. It was our second day in a row diving the *Halfish* and I was finally able to get a few shots of the wreck while Jason Kennedy and Al Pinder both shot video. The next day, Paul Grantham visited the wreck with his own dive boat and shot additional still photos.

Although my camera system was still not working as it should and still seemed to contain a bug, I had finally come home with at least some shots of the wreck.

Jason Kennedy, assisted by Nigel Leggett and Skipper Dave Gray climbs the Ryan & Erin's dive ladder with his video camera after the third Halfish *dive—finally, we have photos of the wreck.*

Author photo.

F.V. Halfish: A Personal Tie 29

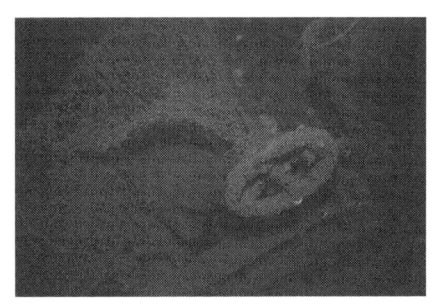

Two shots of the split winches on the main deck. The left shows the two winch drums from the front with their warp wires still evident. The rope tied off on the winch is a semi-permanent buoy line tied in by the author on the fourth dive to the wreck after anchor damage was noted on the third dive. The right photo was taken from behind the starboard winch drum showing the brake hand wheel strung with torn netting.

Paul Grantham photos.

Chapter 4

Forgotten Submarine, HMS L-26

History

On Wednesday, September 2nd, 1998, Swissair flight SR111 took off from JFK airport in New York on a 7 hour flight to Geneva, Switzerland. It was 9:15 pm ADT and no one among the flight's 229 passengers had any reason to think that this trip would be anything more than routine.

At 10:15 the captain sent an emergency message to Boston air traffic control stating that there was smoke in the cockpit. Because Halifax International was only 70 miles away, the decision was made to make an emergency landing there. But first, almost thirty tons of fuel had to be jettisoned from the aircraft to be able to make a safe landing.

At 10:25 Halifax International Airport received a call from SR111 stating that the aircraft was dumping excess fuel and preparing to make an emergency landing. The pilot began a descent from 33,000 feet to 10,000 over the ocean to release the fuel.

At 10:26 as the plane descended it disappeared from air traffic control radar screens in Moncton and Halifax. Six minutes after that, the plane crashed in the ocean near picturesque Peggy's Cove, killing all 229 passengers and the flight crew.

Forgotten Submarine, HMS L-26　　31

In the aftermath of this tragic accident a large scale search of an area encompassing a huge portion of St. Margaret's Bay was conducted. It was hoped that the searchers would find the larger pieces of airplane wreckage and ultimately the cockpit voice recorder. High tech equipment including ROV's, mini submersibles, and underwater camera equipment as well as divers were used to aid in the hunt for the missing plane.

A sonar sweep of the area yielded several targets on the seafloor within the search area. Two of these targets turned out to be shipwrecks. One was positively identified as a small, wooden ship which was probably a fishing schooner and as it was obviously not part of an aircraft it was not investigated. The second showed on the sonar readout as a cylindrical object about 150–200 feet long. Could the plane's fuselage have survived the crash intact? The target was studied further with the aid of an ROV outfitted with camera equipment, but when rusty steel showed up on the video screen it became obvious that this was not part of the plane either. The position of the target was documented, and the target abandoned as the search for the wreckage of the plane continued.

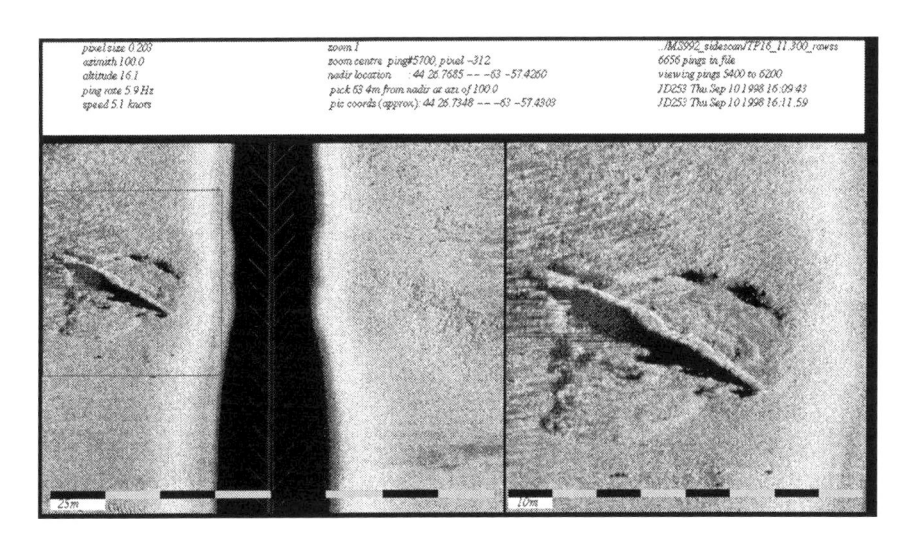

Sidescan sonar image of the L-26 on the bottom of St. Margaret's Bay.

Image courtesy of Gordon Fader.

H.M.S. L-26 *as she looked in the fall of 1937.*

Author's collection.

Word soon spread through the local diving community that a possible submarine had been discovered in St. Margaret's Bay near Peggy's Cove. The small cadre of technical divers in the Halifax area (myself included) was especially champing at the bit as the wreck was beyond the recognized sport diving limit of 130 feet and could perhaps be a virgin. Rapidly the rumour mill inherent in all activities (and in diving as strong as any) had turned the unconfirmed submarine into a World War II German u-boat. As the story spread, divers began voicing the thought, "Could this be another *U-Who* (*U-869*, sunk off the northeast coast of the U.S.)?" At this point, while rumours were flying around the diving community, the crash investigators downplayed the discovery. "There's a 99.9 percent likelihood that it's not a submarine," said Lieutenant-Commander Jim Bradford of Trinity Route Survey. "That won't be determined until Navy divers investigate the scene."

According to retired Chief Petty Officer submariner David Perkins, the target was a British submarine. At the end of the second world war, the decommissioned British L-class submarine *HMS L-26* was acquired by the Canadian government, stripped of serviceable equipment and scuttled several miles off Pennant Point in St. Margaret's Bay to be used as a sonar bottom target for training exercises. "I can recall *HMCS Haida* conducting location and identification exercises on this target in 1956 when I was an Ordinary Seaman TAS rating," Mr. Perkins said. "I understand the target was still in use during the 1960's."

The navy returned to the site after the dust had settled on the Swissair tragedy, after incredibly forgetting about their sonar target for thirty years. More sonar work was conducted and a diver was sent down to the wreck with a video camera to capture images of the site. This work led to the wreck being positively identified as *HMS L-26*.

Due to the success the British had had with their E-class submarines, in 1916 the Admiralty decided to order the construction of two more of these sturdy little vessels taking into account lessons learned from wartime experience and modifying the design accordingly. *E-57* and *E-58* were ordered and built but due to the new modifications and elongated hull it was decided to distinguish these as a new type of submarine and so the L-class was born. These first two submarines of the class were later renamed *L-1* and *L-2*.

A total of sixty-one L-class submarines were ordered, *L-1* thru *L-75*. *L-13* was never ordered due to superstitions surrounding the number 13. *L-37* thru *L-49* were also never ordered. *L-28* thru *L-51* as well as *L-57* thru *L-66, L-70,* and *L-72* thru *L-75* were cancelled before building began or else broken up and sold for scrap before completion. *L-67* and *L-68* were completed in 1927 as the *Harabi* and *Nebojsa* for Yugoslavia.

The L-class submarines can be divided into three types: *L-1* to *L-8* had 18 inch bow and beam torpedo tubes; *L-9* to *L-33*, which were configured as minelayers, with four 21 inch bow tubes and two 18 inch beam tubes; and *L-52* to *L-71* which were fitted with six 21 inch bow tubes.

In addition to their torpedo armament, the L-class boats also carried a gun mounted on the conning tower forward of the bridge. Although the earlier boats (*L-1* to *L-8*) had initially carried a 3 inch gun, all of the boats in the class were eventually fitted with 4 inch guns of various descriptions—resulting in a three-man increase in their complement for the Group 2 boats. This addition meant that the Group 2 boats would carry a 38 man crew. The Group 3 boats which were fitted with 2 four-inch guns would carry 44 men.

The L-class boats were also the first submarines to begin carrying some of their normal fuel stowage in external tanks. Although only twenty tons of fuel was carried in two lightly constructed tanks outside the pressure hull, this was the beginning of the practice of carrying a large amount of the vessel's fuel externally. This method was further refined and came into general usage in the 1920's.

HMS L-26 was built by Vickers at Barrow in Furness and was launched on May 29th, 1919. She was commissioned into the Royal Navy on October 12th, 1926. The submarine was 229 feet in length overall by 13 feet in breadth. Twin 1200 horsepower Vickers diesels turning twin propellers drove her 1080 tons at a top speed of seventeen knots while on the surface. When submerged, she could make a very respectable ten and a half knots on two electric motors producing 1600 horsepower. She could safely dive to depths of 200 feet and could stay at sea for 24 days on the diesels.

During World War II the *L-26* was kept busy as part of the 6th, 5th, and 3rd submarine flotillas, before being transferred to the 7th flotilla to be used as an anti-submarine training vessel. In May of 1944 she was loaned to the Royal Canadian Navy and based in Bermuda until October of that year when she received orders to transfer to Halifax. This was to be her base of operations for the remainder of the war.

As World War II was nearing its end, three of the much older and now obsolete L-class submarines were no longer needed, and in January of 1945 the naval staff suggested that they be turned over to Crown Assets for disposal. These three boats were the *L-23, L-26* and *L-27*. The naval staff in Halifax however, had its own plans for the submarines. They wanted to sink two of them for use as bottom sonar targets for the purpose of training sonar operators, with one to be placed in the Bay of Fundy and the other near Halifax. The Admiralty eventually gave its blessing to the disposal plan and agreed that two of the old L-class subs could be sunk as targets, but in the end only the *L-26* was sunk, in St. Margaret's Bay in September of 1946. The *L-27* was towed to Quebec where she was scrapped, and the *L-23* foundered in deep water while under tow.

Diving the L-26

Although I've now completed several dives on the wreck of *HMS L-26,* there are a couple that stand out. Like most wreck divers, the first time I dive a wreck usually stands out in my mind even after several years, many dives, and many other wrecks have passed. In that respect, the *L-26* is no different.

On the morning of July 20th, 2001, a group of ten divers including myself were at Brenton Gray's boat yard in Sambro, loading our gear onto the *Ryan &*

Erin, Skipper Dave Gray's dive charter boat. That day we would be making a three hour trip to the wreck of the *L-26* in St. Margaret's Bay.

It was a beautiful summer morning, the sun shining and the temperature already in the mid-twenties even at 9:00 am. The wind was calm with the surface of the sea like a pane of glass. We couldn't have wished for better conditions for the long trip to the wreck and all the divers were excited. Many, like me, were making their first trip to the submarine discovered less than three years before.

The long steam to the wreck site seemed to pass quickly for me. While a few of the divers claimed the three bunks in the fo'csle to grab a nap, the rest of us enjoyed the terrific weather and talked about the upcoming dive. Soon, it seemed, we had arrived. While Dave was busy anchoring the *Ryan & Erin* over the wreck, the rest of us were readying our gear. My dive buddy for this dive would be Paul Grantham and we discussed our dive plan while gearing up. We planned for a bottom time of 25 minutes, breathing a 20/30 trimix which would give us a deco time of 40 minutes.

Paul and I were the second team to hit the water. We descended the pale green Poly Steel anchor line towards the wreck. The water in St. Margaret's Bay is often much clearer than that nearer to Halifax and this morning didn't disappoint: the visibility was in the sixty foot range. Shortly after passing the 90 foot mark we sighted the sub. Much of the hull was covered in anemones, making it seem brighter at depth and offering great subject matter for photographers. The sub was lying almost upright on the bottom, with just a very slight port list. We touched down just forward of the conning tower which has toppled off the wreck and lies beside it in the sand. I glanced at my bottom timer, it showed 165 feet. I was amazed to see that although the conning tower had sheared off and her decking was gone, the pressure hull itself was intact. We dropped off the starboard side of the wreck to the bottom at 175 feet and headed toward the bow. Because the submarine is partially buried in the sandy bottom there is only about ten feet of relief from the ocean bottom to the deck.

As we rounded the bow we could see that the torpedo tube outer doors were open. It made the knife edge bow of the sub seem much more menacing somehow. We swam along the port side and back up to the deck as we worked our way aft, once again passing the sheared-off conning tower, noticing an empty binocular box and sound powered telephone attached to the conning tower bulkhead alongside it. Swimming to the stern I thought of how much the design and construction of this sub reminded me of the later Oberon class British submarines. It seemed odd to me that this one was built almost forty years before them.

As we reached the stern and I looked at my bottom timer I saw that twenty-two minutes had passed. I signaled to Paul that it was time to head back to the anchor line to begin our decompression. But not before we took a close look at the twin propellers and after diving planes.

Although much of the decking over the external fuel tanks and inevitable mazes of piping has disintegrated over the years since the sub was scuttled, the pressure hull and even the outer hull are remarkably intact. If not for the missing decking and toppled conning tower she would appear to be simply resting on the bottom, awaiting her captain's orders to begin the next mission. As I ascended from this first, orientation dive on this intriguing wreck, watching her recede into the gloom, I vowed that on subsequent dives I would explore her inner secrets, viewing those spaces where submariners from the 1920's and 30's lived and worked.

Jeffrey Smith swims along the anemone covered starboard bow of HMS L-26.

Mike Grebler photo.

After 43 minutes of decompression obligation I surfaced into the warm July sunshine. What a contrast the heat of the surface was to the 41 degree water on the wreck. As I climbed the ladder to the *Ryan & Erin's* deck, Skipper Dave leaned over the rail and asked, "How was it Al?"

I spit out my regulator, looked up and replied, "What an awesome wreck Dave!"

Certainly it is one of the most awe-inspiring sights to see an intact shipwreck lying on the bottom. Too many of the wrecks lying within diving limits off our shores are little more than scrap heaps, having been pounded flat by storms, salvagers, and explosives. To have a completely intact wreck is a novelty for Nova Scotian divers, and the sight of the 229 feet of the *L-26* is certainly worth the trip for any technical diver, no matter where they're from.

On subsequent dives, Paul Crocker, Jason Kennedy, and I would discover small artifacts and items on the wreck that were missed by a wide-eyed diver on his first dive to an intact submarine. Items such as the attack periscope, with its outer lens and brass cover still intact and protruding from the broken conning tower. Also still attached to the conning tower are the running lights, behind watertight ports, as well as the switches to activate them, housed in ingenious water and pressure tight housings. Although it seems after a first, short dive to this wreck that there's not much to see after you've traveled it's length, upon closer inspection you'll find much to draw your interest and keep you occupied.

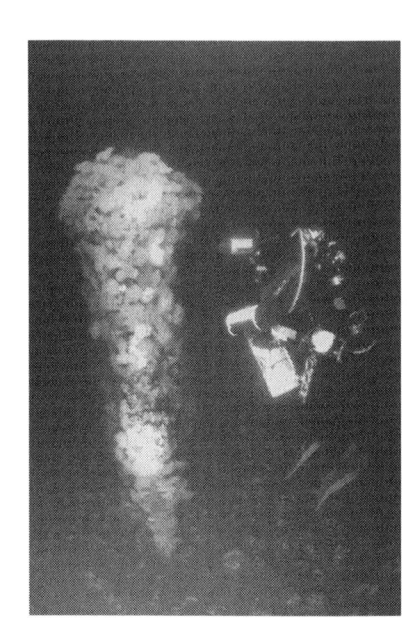

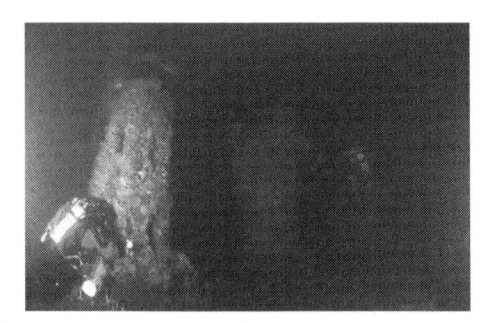

Left: *Jason Kennedy examines the knife-edge bow of the L-26.* **Above right:** *Divers swim by the midships section of the submarine where the conning tower had been located. The upright structures are the hatch/ladderway from the conning tower down into the control room and the periscope wells.* **Below right:** *Jason Kennedy checks out the starboard propeller.*

Mike Grebler photos.

Chapter 5

M.T. British Freedom: Oil on Cold Seas

History

In 1928, at the Tyneside yard of Palmer's Shipbuilding and Ironworks of Jarrow & Hebburn, a new 6984 ton motor tanker was launched for the British Petroleum Company. She was to be named the *British Freedom* and had been born in noble company, for on an adjacent slip, and launched that same summer was the Royal Navy's heavy cruiser *HMS York*. The *York* would be sunk off the coast of Greece in March of 1941 by two Italian EMB's (explosive motor boats), each packed with over half a ton of high explosives. The *British Freedom* would outlive the ship she was born alongside by a scant four years.

However, in that year of 1928 all was well for the rapidly expanding BP Company, as evidenced by the fact that at least ten of the *British Freedom's* sister tankers were all launched that year, all sporting the company's "British" prefix as part of their names.

British Freedom herself was 440 feet in length with a beam of 57 feet and a 34 foot draft when loaded to her marks. Her rigging was typical of the tankers of her time: the machinery spaces and accommodations for the engine room crew were situated aft, with the bridge structure and deck officers and crew being housed in a second superstructure amidships. She was powered by a single 654 horsepower oil-fired steam engine that drove a single gearbox, shaft, and huge propeller which could push her through the water at a maximum of ten knots. At the outbreak of hostilities with Germany she was fitted with a four-inch gun and a smaller anti-aircraft gun on her stern deck thus adding 9 naval gunners to her regular complement.

While steaming along the coast of the United States on June 27th, 1942, the *British Freedom* was torpedoed near Frying Pan Shoals by the Kreigsmarine's Kapitanleutnant Horst Degen in his *U-701*. Although badly damaged with a gaping hole in her cargo tanks, she managed to limp into Norfolk, Virginia. In time, she was repaired and returned to the service of that ever-hungry mistress, the Allied war effort.

Her first trip after this unscheduled refit was to New York, there to load crude oil which she carried across the North Atlantic to the British naval base at Scapa Flow in Scotland. She then spent the next two years hauling highly volatile gasoline up and down the east coast of the United Kingdom.

December of 1944 found the *British Freedom* once again in North American waters and making for a port she knew only too well. On Christmas Eve she docked in New York Harbour, this time to load 9723 tons of U.S. Navy special fuel. After a brutal winter crossing from England with terrible weather including gale force winds, heavy snow, and rain the ship's crew of 48 men and 9 naval gunners were only too happy to spend Christmas in port, even if they couldn't be at home.

After Christmas and the completion of loading, the tanker was to travel to Boston, there to join convoy BX-140 bound for Halifax, where she would in turn join a fast HX convoy bound overseas for England. As it turned out, she missed her convoy and so had to wait for the next one to assemble: the ill-fated BX-141.

The steam tanker British Freedom.

Author's collection.

On January 12th, 1945, at 12:50 pm ADT, the minesweeper escorts *HMCS Westmount* and *HMCS Nipigon* steamed out past the Boston Harbour submarine net gate vessel with twenty merchantmen in a straggling line following astern of them like ducklings swimming in the wake of their mother. Escort Group 27, consisting of the frigates *HMCS Meon, HMCS Ettrick,* and *HMCS Coaticook* would also be providing coverage for the convoy, although Lieutenant R.L.B. Hunter of the *Westmount* would be in overall command.

By 5:45 pm ADT, Captain Frank Llewellyn Morris of the *British Freedom* had jockeyed his ship into its assigned position and the convoy had formed up. Once underway Lieutenant Hunter gave the order for the ships to proceed at 7.5 knots on a course for Halifax. This signal was in turn relayed by the Convoy Commodore and Rear Commodore to all of the merchant ships (it should be noted that the Convoy Commodore would usually be a retired naval officer with his own staff who shipped aboard a suitable merchant vessel for the voyage).

The ships of the convoy and their escorts had an uneventful passage until the morning of the 14th when they were preparing to enter Halifax Harbour.

As the ships were forming into a single line in preparation for entering harbour through the anti-submarine nets at Meagher's Beach, little did they expect that they were being closely watched. Kapitan zur See Kurt Dobratz in the

U-1232 was at periscope depth nearby and keenly scrutinizing the assembled ships. Up to this point Dobratz had had a frustrating cruise, with the bitter cold weather freezing up his periscope and schnorkel apparatus time and time again and making them nearly useless. To make matters worse for his crew, they had originally been sent to the middle of the North Atlantic to act as a weather ship, periodically sending radio messages of prevailing conditions back to U-boat Command in Germany. But their perseverance and persistence, it seemed, was about to pay off, for at the end of their weather watching vigil Dobratz had headed for the eastern coast of Canada with hopes of milder weather and good hunting. On the morning of January 14th, 1945 he had found his prey.

The crews of both the merchantmen and the escorts of BX-141 were just beginning to relax at the end of what seemed at that point to be just a routine run when an explosion split the morning stillness. An acoustic torpedo from the *U-1232* had hit the third ship in line in the engine room on her starboard side. That ship was the *British Freedom*. The time was 10:32 am. The next vessel in line was the Rear Commodore's ship *Martin Van Buren*. Seeing the explosion from the bridge of his Liberty Ship, Captain Hiss of the *Van Buren* gave orders to pass to starboard of the stricken tanker. At 10:40 am as she passed alongside the *Freedom*, the *U-1232's* second torpedo took the *Martin Van Buren* in the stern. This explosion not only killed three of her four naval gunners but also left the ship dead in the water. She later drifted ashore on the rocks near Sambro and became a total loss. Her bones now lie in 50 feet of water and the wreck is a frequent destination for sport divers.

With these two taken care of Dobratz turned his attention to the seventh ship in line, the large molasses tanker *Athelviking*. The tanker's master, Captain Martin, having witnessed the fate to befall his two compatriots, altered course to pass down the *British Freedom's* port side. This worked for him no better than passing the *Freedom's* starboard side had worked for Captain Hiss and at 10:45 a third torpedo from the u-boat hit the *Athelviking* in the stern. She later sank in 300 feet of water some distance to the southwest of the *British Freedom* and the site where the attack occurred.

The first to send a distress message was the *Martin Van Buren*. The signal "SSSS", meaning that the ship was being attacked by a submarine, flew threw the ether. This was quickly followed by the distress call of the *British Freedom* at 10:57, with the *Athelviking's* call coming two minutes later.

Captain Morris had remained aboard his ship with a few of his men after ordering the rest of the crew to take to the boats. Perhaps he still had some hope of saving his ship and getting her safely into harbour. If so, his optimism was little more than a pipe dream. The *British Freedom* was settling quickly by the stern and it was clear to all those still aboard that she was doomed. The tanker's chief radio operator, Bert Hawling, had radioed for a tug to come and take them in tow after he sent his distress call, but even had one put out from Halifax it would have arrived too late to save the ship. By the time the captain and remaining crew abandoned ship in the last remaining lifeboat and a raft, the tanker's entire after end including her funnel was underwater.

Meanwhile, Dobratz wasn't being let off scot free for his impertinence. The convoy escorts were searching for the *U-1232* like angry hornets whose nest had been disturbed. During the search for the submarine *HMCS Ettrick* got an asdic contact and ran down upon the u-boat. Dobratz ordered a crash dive but was just a little too late: as she was diving the *Ettrick* rammed into her conning tower, destroying the bridge as well as the periscopes and carrying away the radio antennas. Nevertheless, the integrity of the pressure hull wasn't breached and Dobratz managed to slink away to a deep trench off the Sambro Ledges, where he and his crew waited silently on the bottom in suspense for more than two days while destroyers continued the search for the impertinent one. He had torpedoed and destroyed three ships in just thirteen minutes, an achievement for which he would be awarded the Knight's Cross when he returned to Germany. Both he and the *U-1232* would survive the war. That wasn't to be for one of the *Freedom's* crew.

The ship's electrician, thirty-three year old W. Henderson, was killed instantly in the initial blast. He had been in the engine room when the torpedo struck and would have had no chance to do anything to save himself. The fourth engineer was on watch at the time but was in the boiler room when the torpedo struck and was temporarily blinded but luckily not killed. He was led on deck to safety by the oiler who had been on watch with him.

With the exception of Henderson, the entire crew of the *British Freedom* as well as the ship's dog got away in the lifeboats and raft. They were picked up a short time later by the minesweeper *HMCS Gaspe*. An excerpt from the *Gaspe's* log on that morning reads:

1120—British Freedom sinking. Approximately 1000 yds. From # 1 buoy. Standing by to pick up survivors. Guns crews closed up. Action stations.

1125—Picked up first lifeboat.

1133—Picked up second lifeboat.

1135—Picked up third lifeboat.

1143—Picked up fourth lifeboat.

1147—Picked up raft. Total survivors picked up from British Freedom: *56. Lifeboats cut adrift. Proceeded into harbour.*

*1202—*British Freedom *down stern first, only fo'csle above water.*

The *Freedom* remained in that sorry condition until the next morning, with her stern resting on the bottom and her bow still above the surface and clinging desperately to life. On the morning of the 15th, the Bangor class minesweeper *HMCS Goderich* was sent out of Halifax to depth charge and sink the *British Freedom*. She was still partially afloat but there was no chance of liberating her from her grave, and as she had gone down in the busy shipping lanes she had become a hazard to the safe navigation of other vessels and so had to be sunk.

Diving the British Freedom

The *British Freedom* was rediscovered in 1995 during the same survey by the Geological Survey of Canada and led by Gordon Fader which was mentioned in the chapter about the *Halfish*. One of the targets in the approaches to Halifax Harbour which showed on the images was believed to be the wreck of the *British Freedom*. In July of the same year, identity of the wreck was confirmed by Canadian Navy divers who visited the wreck in a submersible launched from the support ship *HMCS Cormorant*. Sport divers weren't long in following the Navy's lead.

A group of divers operating out of the now defunct Discover Diving Scuba Shop in Bedford were the first to dive the wreck later in the fall. Nova Scotia's technical diving community was still in it's infancy at the time, local dive shops being still wary of embracing the new mixed gas technology that the wreck divers in the northeastern U. S. were at that time taking advantage of to make their deep-diving activities safer. For this reason most of the early dives to the wreck were conducted employing short bottom times on air with surface supplied oxygen on a deco stage for decompression. This was to become the norm for deep exploration in the area for the next few years until mixed gas training became readily available in Nova Scotia.

The *British Freedom* now lies on the bottom in 200 feet of water just east of the outbound shipping lane from Halifax Harbour. There are two main pieces of intact wreckage for divers to visit: the stern deck area from the taffrail to the boiler room area and a section of the forward deck. Joining these two sections is a large debris field of broken and scattered hull and deck plating. Beyond the forward deck section is another, much smaller debris field which includes the very tip of the bow. Although it is possible to swim the entire wreck during a dive, a free-swimming diver without a DPV (diver propulsion vehicle or scooter) doesn't have much time to stop for a close look at things. For this reason, the wreck is usually split into two separate dives, bow and stern, each of which is normally buoyed during the summer diving season.

Descending to the stern of the *British Freedom* the diver will land on her afterdeck at 160 feet. Until Hurricane Juan struck the area in the fall of 2003, this deck was sitting level and sporting the remains of the after accommodations as well as the four inch stern gun and anti-aircraft gun in it's turret (which was so well greased that even after being submerged for almost sixty years it could still be trained and elevated by divers cranking the hand wheels). After the hurricane hit, this area was found to be collapsed to starboard and now sits with the port side at 160 feet while the starboard side of the deck is at 190 feet. Although the starboard side of the hull is still standing as a vertical wall above the collapsed deck, the AA turret took on a precarious forty-five degree list over the starboard rail and the following year it collapsed into the sand, crushing part of the hull as it went.

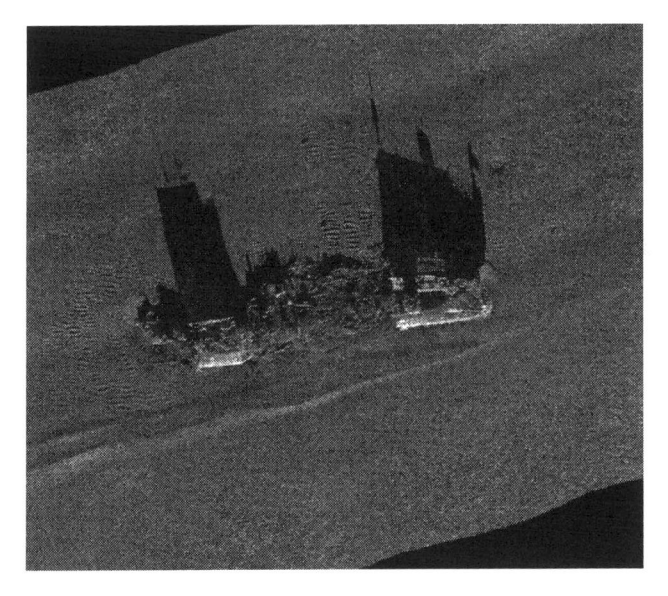

A side-scan image of the British Freedom *as she was when rediscovered. The two main pieces of wreckage can be easily discerned, as well as the shadows of the stern gun (far right) and anti-aircraft turret with its gun barrel pointed upwards.*

Image courtesy of Gordon Fader, GSC.

Amazingly, one of the *British Freedom's* most intriguing features wasn't destroyed by this second collapse. Most visitors to the wreck are excited at the prospect of swimming through the massive torpedo hole in the engine room's starboard side at 190 feet. From there you can swim through the remains of the engine room and back up to the main deck at 160 feet. Sadly, there is little opportunity for penetration in this area now because of the main deck collapse. Nevertheless, there is still a picturesque swim through the torpedo hole from outside the hull, then up past the boilers, passing crates of spare engine parts and dislocated ammunition for the AA gun. It is still possible for divers to access the port side of the boiler room, where numerous catwalks, ladders, pumps, and other equipment can be seen. In July of 2006 I did a dive on the stern section with Greg Mossfeldt and found that swimming through this area of the boiler room as well as the torpedo hole in the hull still gives me as much of a thrill as it did on my first *Freedom* dive.

Access to the port side of the after accommodation area under the main deck is also still possible although the starboard side has now collapsed. This is the area where the stokers, oilers, and other engine room personnel would have made their home. The above deck accommodation for the engine room officers and cooks has sadly disintegrated and collapsed. In addition to the after accommodation area, in this space is also to be seen the remains of the after emergency steering station with its compass binnacle still in situ. Numerous artifacts such as ship's china and 78 rpm records have also been found in this area.

Until the collapse much of the original teak decking was still in place as well, although now most of it lies scattered and splintered about the area. In this area also can be found numerous three inch and four inch shells for the guns, some still in their original wooden crates, others strewn loose around the stern area.

Near the forward end of the stern section can be found the remnants of the galley sitting atop the remains of the main deck, with the large, cast-iron cook stove still in residence. This area of the wreck provides magnificent opportunities for exploration and photography.

The anemone-covered four-inch gun points over the stern rail of the tanker British Freedom. **Inset:** *The author with the stern gun's telescopic sight. This was recovered from the opposite side of the gun to that shown in the photo.*

Mike Grebler photo.

Swimming forward of the boilers past the point where the stern section is broken off, the diver will pass over a huge, twisted mass of fuel transfer hoses which then give way to the more common shipwreck debris of tangled steel beams, ribs, and plating as the debris field proper is reached. Near the forward end of the debris field is the collapsed wreckage of the forward superstructure, now blended in with the hull wreckage and identifiable only by the splintered teak decking and numerous portholes in attendance. Several portholes were recovered from this area lying loose in the sand to port and starboard of the wreck during the early days of exploring the wreck.

The debris field portion of the wreck has still been little explored although Paul Crocker and I spent many of our *Freedom* dives during the 2005 and 2006 diving seasons in this area. After beginning to explore the massive debris field on the *Kolkhosnik* with Jason Kennedy (see Chapter VI) we had decided to do the same on the *British Freedom*, rather than confine our dives to the intact portions of the wreck. This led to us not only being able to navigate the wreck better by landmarking, but also led to discoveries of artifacts and features that we had previously had no idea existed.

A cargo bulkhead towering thirty feet above the bottom wreckage marks the end of the debris field and as you slowly ascend and rise above it you are greeted by the sight of the intact forward deck at 155 feet. Here can be seen a spare anchor still lashed to the deck, line-handling winches, and oil transfer piping and valves. A large hatch in the deck provides access to the only remaining intact cargo tank, dropping down about forty-five feet to below the level of the sand outside. This intact area is the full width of the ship and approximately 70 feet of her 440 foot length.

Continuing forward you come to the collapsed bow section, sitting at 190 feet. This area is marked by the ships huge anchors and chains, the remains of her chain locker, and the forepeak. Beyond the bow lies the almost flat sand and gravel bottom that the *British Freedom* came to rest upon, with her bow pointing to the west-northwest and the land that she would never reach.

The wreck of the *British Freedom* is a deep, dark dive that is significantly brightened by the thousands of sea anemones that seem to cover every upright surface on the ship. When descending, divers will know they are nearing the wreck as the white tentacles of these creatures cast a subtle glow in the gloom. The abundance of anemones can make it difficult to discern individual features for what they truly are, but for the experienced shipwreck explorer, this only adds to the intrigue of this amazing wreck. Whether you're a seasoned deep-wreck diver or a neophyte to the world of technical diving, the *British Freedom* is a must-dive shipwreck. Like me, once you've made that first dive to the tanker, you'll be hooked.

CHAPTER 6

S.S. Kolkhosnik: aka The Russian

History

In 1904, John Readhead and his five sons who were already partners in the company of John Readhead & Sons Ltd. Shipbuilding of South Shields, England, decided to form a new company to boost their dwindling shipbuilding business during the recession which followed the Boer War. They formed the Cliffe Steamship Line for which they built their own cargo vessels to carry their freight (a smaller version of K.C. Irving's vision later in the 20th century).

Beginning in July, 1904 with the launching of the first *Rockcliffe*, the company went on to build seven ships which they operated until the company's voluntary liquidation in May of 1941. All of their vessels, with the exception of the *Ulidia* which was a Readhead built ship purchased from the Mercantile Steamship Company of Ireland, sported the "Cliffe" suffix in their name.

The original *Rockcliffe* had been sunk by German gunfire in 1916, so when a new ship for the company was launched on March 26th, 1925, she was christened *Rockcliffe*, official number 139915.

All the Cliffe Line ships were very similar in design, size, and outfitting. This new *Rockcliffe* fit the mold and was virtually indistinguishable from her sister ships. She was 3880 gross registered tons, 364 feet in length, and 51 feet in breadth. She drew 24 feet of water when loaded and could make 10.5 knots at full power. Her power plant was comprised of twin Scotch boilers driving a huge triple expansion steam engine and single 144 inch three bladed propeller. She was designed in the manner of most contemporary cargo vessels with a high fo'csle and after castle and a raised center castle sporting the superstructure and machinery spaces. Forward and aft of the center castle were the cargo holds.

The Highcliffe, *one of the Cliffe Line's sister ships of the* Kolkhosnik.

Author's collection.

After the Great Depression, the entire fleet of Cliffe ships (comprising four at that time) was laid up. The *Rockcliffe* herself tied up in Shields on October 20th, 1930 and was not to move again until she was sold in July of 1935.

She was purchased by the Russian Sovtorgflot (Soviet Commercial Fleet) and renamed the *Kolkhosnik,* an appropriately communist name which means "collective farm worker".

At the outbreak of the Second World War, the *Kolkhosnik* was requisitioned for war service and was put on convoy duty hauling war cargoes from North America to the Russian ports of Archangel and Murmansk. She was armed with two World War I vintage three inch guns, one on the bow and the other at her stern, and thus equipped was sent back to sea to ply her trade in the u-boat infested waters of the North Atlantic.

On December 11th, 1941 the *Kolkhosnik* entered Boston Harbour to load war cargo for Russia. The Japanese had four days earlier attacked Pearl Harbour and the Americans had just entered the war. After a long wait to finish loading her freight, the *Kolkhosnik* finally left Boston on January 15th, 1942, sailing alone up the coast to Halifax to join Convoy SC-65 to Liverpool, England, there to join a convoy bound for Archangel. She was destined never to reach the Nova Scotian port.

Because u-boats were at the time extremely active along the coast of North America, Captain George Zarev decided to keep his ship as close inshore as possible, hoping in this way to steer clear of any German raiders who might have been in the area. It seems he strayed a little too close to shore while off Sambro and at 1:25 am on January 16th the ship struck and was holed by Smithson Rock.

Although the ship didn't slow down until Zarev ordered the engines stopped, she was making water very quickly in number 2 cargo hold which swiftly found it's way into the stokehold and boiler room and the crew barely had time to take to the lifeboats before she sank, carrying one of her crew with her. An engineer would later die in one of the lifeboats on the way to Sambro Island. Zarev later reported that the *Kolkhosnik* went down in just 25 minutes.

Even though it's almost certain that the *Kolkhosnik* went down as the direct result of a brush with Smithson Rock, the captain and crew later adamantly insisted that they had been torpedoed, stating that because the ship lost no forward way they couldn't have struck a rock. The Department of External Affairs in Ottawa also concluded that the ship was lost due to u-boat activity.

There is good reason to believe that the *Kolkhosnik* was not, in fact torpedoed. The crew stated that she was hit on the port side; this would have meant that the u-boat who torpedoed the ship would have had to be inshore of her, among the dangerous rocks and shallow water in the approaches to Sambro. No self-respecting submarine commander worth his salt would have placed his vessel in such a dangerous and precarious position. Furthermore, if it had been concluded that the *Kolkhosnik* had sunk due to contacting Smithson Rock, there would have almost certainly been an inquiry into the circumstances of the loss and a good chance that Captain Zarev would have had his master's ticket suspended. He then, had good reason to make the investigating board believe that a torpedo had sunk his vessel.

After the war the British Company of Risdon Beazley Limited had contracted to recover the cargoes of merchantmen sunk during the war. One of these vessels was the *Empire Kingfisher* (see Chapter X) which had sunk on Nova Scotia's south shore. Risdon Beazley's operation essentially consisted of putting a man down in a one-atmosphere bell, who then directed the topside crew in where to lay explosive charges to open up the wreck to gain access to the cargo. When this was done, the man in the bell then guided an eight-legged grapple from the salvage ship's derrick which was used to recover the ship's cargo.

In 1951, a local company was attempting to recover the *Kolkhosnik's* cargo of zinc and nickel using hard-hat divers deployed from a surface support vessel. There were several problems with this scenario, the chief of them being two other portions of the *Kolkhosnik's* cargo: barbed wire and tetra ethyl lead.

Tetra ethyl lead is the additive that used to be mixed into gasoline to prevent engine knocking. It is a highly toxic, heavy liquid that will not mix with water. On the *Kolkhosnik*, this part of the cargo had been carried in fifty-gallon drums, many of which had split open and spilled their contents by 1951. This resulted in large pools of tetra ethyl lead lying on the sea floor in the vicinity of as well as inside of the wreck.

The hard-hat divers, while working in poor visibility, would inevitably tear their canvas suits on the barbed wire, which then allowed the tetra ethyl lead access to their skin. Being absorbed in this way and causing such violent migraine headaches that the divers could no longer work, the tetra ethyl lead made the wreck unworkable and the job was given up as impossible.

Impossible for divers perhaps, but not for Risdon Beazley's crew of expert salvors. When work on the *Empire Kingfisher* had concluded in 1952 the company's salvage vessel *Help* headed for the site of the *Kolkhosnik* and successfully salvaged 1062 tons of her cargo of zinc, nickel, and molybdenum.

The *Kolkhosnik* was then forgotten until the early 1970's when another company decided to have a go at the wreck's cargo. Risdon Beazley's methods typically left about fifteen percent of a wreck's cargo on the bottom, and armed with this information these new salvagers attacked the wreck with a will, utilizing divers in scuba gear to hand-pick the remaining nickel and zinc cargo, which was then placed in drums and winched to the surface. By this time the tetra ethyl lead had dispersed and so was not a problem for the divers.

By the mid-1980's, sport divers had discovered the *Kolkhosnik*, or "The Russian" as it had been dubbed by the area's fishermen. One of the early sport diving trips to the wreck was led by Gary Gentile and a group of American and local wreck divers. Gary wrote of diving the *Kolkhosnik* in his book Wreck Diving Adventures.

Ten years later when I first visited the wreck, it was still infrequently visited. Being at the limits of the certifying agencies recommended depth for sport divers, it was a difficult site to get to unless you had your own boat because the local dive shops and charter operators were fearful of liability.

As we moved into the new millennium attitudes and training had changed drastically, and the wreck is now a "must see" destination for up and coming technical divers, technical dive training, and of course still a favorite with those of us to whom "The Russian" has become a familiar, old friend. In fact, diving the *Kolkhosnik* has become so popular of late that Skipper Dave's Charters of Sambro ran over twenty charters in 2006 to this wreck alone. That's in addition to the numerous other wrecks that Dave visits during our short dive season.

Diving the Kolkhosnik

The *Kolkhosnik's* bones now lie on a granite boulder and sand bottom 0.8 nautical miles east of Smithson Rock and 2.2 nautical miles south-southwest of Sambro Island light. She sits on a roughly north-northeast/south-southwest axis with her bow pointing towards Sambro Island, which Captain Zarev steered for after the ship struck the rock and began taking on water. Her bow rests in 120 feet of cold North Atlantic water with her stern somewhat deeper at 150 feet.

Other than at her boilers and a small intact portion of the bow, which rise to 120 and 110 feet respectively, there is little relief on the wreck. Risdon Beazley's crew did their work well: almost the entire wreck has been blown open, flattened, and scattered. Although surrounded by a large debris field, it is possible to follow the line of the ship's starboard side from bow to stern along the explosives cut at the turn of the bilge. The vessel's double bottoms are mostly intact below where the cargo would have been stowed, and these create a large, flat expanse of steel both forward and aft of the boilers and engine.

As one swims along the wreck, several Grant tanks can be seen. These twenty-six ton monsters were part of the *Kolkhosnik's* deck cargo and make fabulous subjects for photographers. The tanks were powered by two 95 horsepower Leyland diesel engines and were bound for the Russian front during that winter of 1942.

The propeller shaft still rests on its bearing beds, reaching from the engine to the collapsed after deck, where it disappears under wreckage, only to reappear where the iron propeller itself is attached. Also at the stern is the after three-inch deck gun which, with the exception of the tanks, is surely the most photographed feature on the wreck. This reminder of the war is still in its mount above the stern, elevated towards the sky and seeming still vigilant.

The engine and boiler room areas are a nightmare of wiring, copper and steel pipes, and other entanglement hazards, with the massive triple expansion engine and twin scotch boilers towering over all. However, it is possible to closely explore this area if one is careful to avoid the perils.

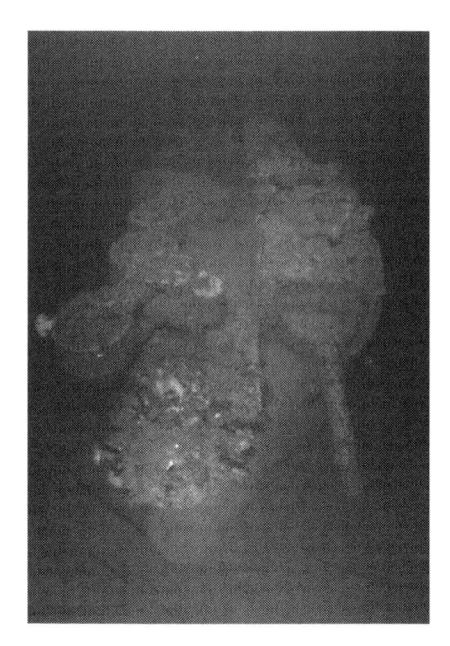

Two of the Kolkhosnik's *deck cargo of twenty-six ton, Leyland diesel powered Grant tanks.*

Mike Grebler photos.

About the only remotely intact portion of the wreck is the bow. The area of the fo'csle and chain locker is laying on its starboard side, festooned with fire hoses, shells for the three-inch bow gun, and scattered cargo and other debris. The bow gun, unlike its intact aft counterpart, lies partially buried in the sand to starboard, with wreckage tumbled across it. The gun's mount is resting upright on the bottom some distance away.

East of the bow can be found a spare propeller, lying on the rocky bottom some distance from the wreck. Also in this area are numerous truck and jeep parts, batteries, and of course the ever present barbed wire and copper telephone wire. Here can also be found two rotary airplane engines, discarded in the sand by Risdon Beazley's eight-legged claw.

My first excursion to the *Kolkhosnik* was on June 22nd, 1996. Commercial diver Dave Pilot had recently purchased a 29 foot Tancook Island built boat which we had refitted into a dive boat and he had christened the *Nova Sea*. One of our first trips in this vessel was to the *Russian*. Dave and his buddy Duane Williams, along with me and my longtime friend Pete DeGrace headed out on this sunny Saturday morning from our home port of Eastern Passage for a dive that would begin for me a long-time love affair with this wreck.

Diver Steve Crain sights along the barrel of the three-inch stern gun at 140 feet.

Mike Grebler photo.

In just under two hours (compared to the 20 minutes it takes to get there from Skipper Dave Gray's wharf in Sambro today) we had arrived at the wreck site to find that it had been thoughtfully buoyed earlier in the month by charter operator Tony Gillis. We tied the boat to Tony's down line and began readying our gear for the dive. While gearing up, Duane remarked that he felt ill and wasn't going to dive and Dave also elected to stay on the boat. They assisted me and Pete in gearing up and we then splashed in.

Descending the line, we found that the visibility was phenomenal and as we approached 60 feet we could see the wreck opening up beneath us. I could see hundreds of circular objects bunched together in the wreckage beneath me which looked to be about the size of coffee cups but I couldn't determine what these might be. It was only as we got closer to the wreck and these circular objects kept getting larger that I realized they were rolls of barbed wire about two feet across:

the visibility was so good that I had been able to discern the individual coils while still sixty or so feet above the wreck.

We touched down on the forward part of the wreck in 130 feet to find that the down line had been tied to the gun barrel of a Grant Tank, once part of the *Kolkhosnik's* deck cargo and now lying on its side and missing its tracks. This was going to prove to be one awesome dive!

Left: *Paul Crocker swims over a Grant tank lying on its side in the wreckage.* **Right:** *Two-thirds of a zinc ingot from the* Russian's *cargo of raw metals for the war effort. The writing on the complete ingots reads "Anaconda High Grade". At the lower right of the photo can just be made out a smaller square object: this is a two-pound nickel ingot, the valuable metal that led to the wreck being salvaged.*

Mike Grebler photos.

We started swimming aft along the midline of the wreck, passing by another tank and soon coming to the two huge boilers which tower over the wreck site. After swimming around the boilers on the port side we arrived at the main engine and soon found the shaft leading towards the stern. We followed the shaft to the almost flattened after castle, rising off the bottom for a close look at the stern gun, still standing and pointing towards the sky as if towards an unseen enemy. We dropped back down to the sand for a quick look at the propeller in 150 feet and then turned back towards the bow.

After once more passing the boilers, we took in the remnants of the *Kolkhosnik's* cargo of war materials: There were hundreds of coils of barbed wire and field telephone wire as well as the field phones themselves. Airplane engines, jeep tires, radiators, truck axles, and other vehicle parts were strewn about the site as if tossed there by a giant hand. Of course the most remarkable and memorable part of her cargo I've already mentioned: the deck load of Grant Tanks of which at least five were in evidence. One was lying on its side, another completely flipped over on its turret, and a couple still sitting upright, their rubber treads (as opposed to steel tracks) testifying to their intended destination and use in Russia during the winter time.

Jason Kennedy swims along the partially collapsed bow section of the wreck. This is one of the shallowest parts of the Kolkhosnik, *reaching to within 115 feet of the surface.*

Mike Grebler photo.

In this forward portion as well, it is easy to see the cut line just above the turn of the ship's bilge, where Risdon Beazley's explosives did their job in opening up the wreck to allow the grapple access to the cargo holds. The cuts are so clean they look as if they were made with a can opener.

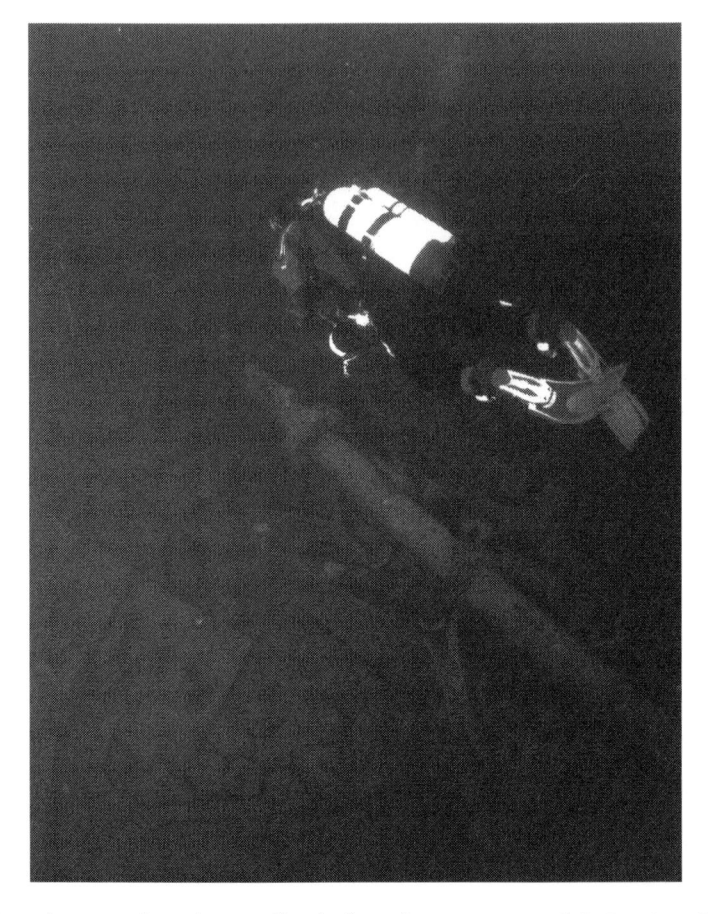

Jason Kennedy swims along the propeller shaft on the stern section of the Russian. *Beneath him can be seen coils of barbed wire from the cargo holds.*

Mike Grebler photo.

All too soon it was time for us to leave. We had been almost thirty minutes on the wreck and had about 40 minutes of decompression obligation according to my dive computer. We ascended to our first stop at 60 feet, following our computer's profiles and ascending by ten foot increments as we slowly off gassed nitrogen from our bodies. At the thirty foot stop, while adjusting some gear, Pete accidentally unclipped his back up computer from it's D-ring and we watched it quickly sink out of sight (This computer was an old Hans Hass Deco-Brain, one of the first dive computers, which he had purchased from me and was so large that Pete elected to clip it on his BCD rather than strap it to his wrist). On a subsequent dive to the wreck we quickly located the computer in the sand well off the wreck, which, when recharged, worked flawlessly even after having spent more than a week on the bottom. As of this writing, Pete still has it.

After completing our lengthy decompression, we surfaced into a beautiful early summer day. When we had climbed aboard the boat and removed our gear, we both agreed that it had been our greatest dive to date. As I noted in my dive log following the dive that day: "Dave and Duane didn't dive today—Sucks to be them!"

Since that first dive Pete and I made on the *Kolkhosnik*, I've visited the wreck over 50 times and find there is always something new to see or a new place to explore. I've since learned the wreck well, to the point where I can be dropped by the dive boat on almost any part of the wreck site and know instantly where I am. On one of my earlier dives while exploring the debris field, an area that I didn't know well at the time, I happened upon a small, thin piece of brass sticking out from under some debris. Upon closer inspection I realized two things: the first was that what I had noticed were the ship's brass parallel rulers, displaced from the chartroom and lost in this area away from the main wreckage. The second was that the debris they were trapped under was a large piece of tank track. By carefully wiggling and tugging the rulers, I managed after a few minutes of work to free them from their tomb. I had just recovered a fabulous memento of this fascinating wreck. After careful preservation and restoration this artifact would hold pride of place in my study.

In the summer of 2005, Jason Kennedy, Paul Crocker, and I began systematically exploring this large debris field to the east or starboard side of the wreck. We continued these explorations during the 2006 dive season. This area of the wreck is where materials unwanted by Risdon Beazley's salvage crew were dumped to get them away from the wreck and out of the way of the valuable cargo and until now has been little investigated. There have been some interesting discoveries made in this area, including yet another Grant tank, lying upside down with it's turret nearby, which Jason discovered in 2005. Discoveries like this only reinforce the theory that there is still much to see and be discovered on this amazing World War II wreck.

The author with the set of brass parallel rulers recovered from the Kolkhosnik. *These instruments were used when laying off the ship's projected course on nautical charts.*

Photo by Paul Crocker.

Chapter 7

▼

Collision at Sea: M.S. Kaaparen

History

Hull number 435 at A/B Gotaverken in Gothenburg, Sweden was launched on May 8th, 1930 for the Norwegian America Line. The ship was 362'1" in length by 50 feet wide and displaced 3156 gross tons. Like her sister the *Tonsbergfjord*, launched from the same yard two months previously, her name was to incorporate the suffix "fjord", a long standing tradition with the company. She was christened *Larviksfjord* for an inlet in Norway, although she would keep that name for only a short time.

After several transatlantic trips, the *Larviksfjord* caught fire near Sweden on September 26th, 1931. When the blaze in her engine room could not be brought under control, her captain ordered his crew to abandon ship and the vessel was left drifting and unmanned. The tug *Helios* found her in this condition and managed with superhuman effort to get a tow line connected to the burning ship. The *Larviksfjord* was taken in tow but her trials weren't over yet. Just outside Stockholm she ran aground, there to sit, still burning, for the next three months. At the end of December the *Larviksfjord's* funeral pyre was finally extinguished. By this time she was little more than a blackened, blistered shell and was condemned.

Still, a Swedish company belonging to G. Carlsson, Rederi Transatlantic, thought there might be some use in the gutted hull and the company purchased the derelict. After negotiating with A/B Gotaverken to rebuild the ship, in the summer of 1932 she was towed ignominiously back to the yard where she had been launched just two years before.

The burned out hulk was given the new hull number 466 and work was immediately commenced to get the ship back into sea going condition. She was completely refitted and furnished with refrigeration equipment in her cargo holds, thus making her well qualified to be put on the South Atlantic run, hauling fresh fruit back to Sweden from South Africa. She would be the first Swedish vessel to do so.

On May 13th, 1933, just three years and five days after she was first launched, the ship was born again and delivered to her new masters. This time her given name was *Kaaparen*, official number 7901.

This new vessel built from the ruins of the old displaced 3385 tons. She was powered by two 8 cylinder B&W Diesels of 724 hp each turning a bronze propeller which could thrust her through the water at speeds of up to 16 knots. Beneath her decks were five cargo holds, three forward and two aft of her midships superstructure, with numbers 2 & 3 being the ones fitted with refrigeration gear. The ship could also boast four diesel generators to provide power, air and refrigeration compressors, and ten winches for handling cargo.

In addition to being well outfitted mechanically, the *Kaaparen* was also spacious in regard to living quarters. Accommodations were available for 42 crew in 32 cabins as well as 12 passengers in 6 double berth cabins. *Kaaparen* also sported a large, modern galley, mess room, and saloon for the comfort and culinary enjoyment of passengers and crew.

In her new incarnation, the ship passed the rest of the decade uneventfully, at least until Poland was invaded and war was declared. She was then put on the convoy run, supplying the insatiable Allied war machine with materials sent across the Atlantic from North America.

The twin screw passenger and cargo vessel M.S. Kaaparen.

Paul Grantham collection.

While the old hoodoo from the ship's days as *Larviksfjord* may have been resting, it wasn't yet ready to give up completely on the *Kaaparen*. On April 27th, 1942 while in convoy from England to Halifax, the *Kaaparen* was involved in a collision with the *M.S. Explorer*, resulting in severe damage to both vessels. The *Kaaparen* was repaired in time to be loaded in Halifax with a cargo of nickel and aluminum as well as frozen meat in holds 2 & 3 and to join Convoy HX 194 bound for the United Kingdom.

At 10:58 am on June 14th, 1942 the first ship in the convoy of 32 vessels got underway from Bedford Basin. The ships followed single file down the Harbour and out through the anti-submarine nets. *Kaaparen* was the thirteenth ship in line, a number that would prove very unlucky indeed in the hours to come.

Once clear of the land south of Chebucto Head, the order was given for the convoy to form up into its columns. *Kaaparen* had been assigned to position 23 at the convoy meeting the previous day, which meant she would be the third ship in the second of the nine columns of this convoy. The ship assigned to position 13 on her port hand was the 5500 ton Norwegian freighter *Tungsha*. At 3:15 pm the *Tungsha* was approaching the gathering convoy from astern to take up her position. As she came near the *Kaaparen*, one or both vessels altered course causing the *Tungsha* to ram the Swedish vessel in her port side. After the dust settled the *Tungsha* returned to Halifax for repairs and left on her postponed voyage a week later as part of Convoy HX-195.

The *Kaaparen* didn't fare so well. Despite the heroic efforts of *Kaaparen's* crew, the *Tungsha* had accomplished what the *Explorer* hadn't been able to do two months earlier. At 9:03 pm on June 14th, 1942 the Swedish motor ship *Kaaparen* gave up the struggle and was engulfed by the salty Atlantic.

In 1952, the firm of Risdon Beazley launched a brand new salvage vessel called the *Twyford*, the first ship in the world to be purposely built for deep-water cargo recovery. By 1953, the *Twyford* was on the east coast of Canada, replacing the tug *Help* which had salvaged the cargoes from the *Empire Kingfisher* and the *Kolkhosnik*. Besides raising the sunken ammunition ship *Trongate* in Halifax Harbour as a government contract, the company worked two further shipwrecks in Nova Scotia's waters for their precious raw-metal cargoes: the *Alexander Macomb* on Georges Bank and in 1955–56, the *Kaaparen* near Halifax.

Unlike the *Kolkhosnik*, the *Kaaparen's* exact location wasn't known. The Navy position for the sinking was taken by land bearings and because of that couldn't be completely accurate, which meant the *Twyford's* skipper had to search a grid with sonar to locate the wreck, using the Navy's position as the center of his search area.

Once the wreck was found, it was business as usual for the salvage gang. First, six mooring blocks, their chains and buoys were laid out to moor the salvage ship above the wreck. Once the *Twyford* was moored over the wreck the "diver" was sent down in his bell from which he directed the laying of charges to blow open the wreck's cargo holds.

After the explosion's aftermath had cleared, the diver was again sent down to the wreck to coordinate the recovery of the cargo using the eight-legged, four-ton grab attached to the salvage ship's derrick. In the end, the *Twyford* succeeded in recovering 350 tons of nickel and aluminum ingots from the shattered hulk of the *Kaaparen*. The *Twyford* recovered her moorings and steamed away to other jobs, leaving the *Kaaparen*, broken and forgotten, to slowly rot in her watery grave.

Collision at Sea: M.S. Kaaparen 65

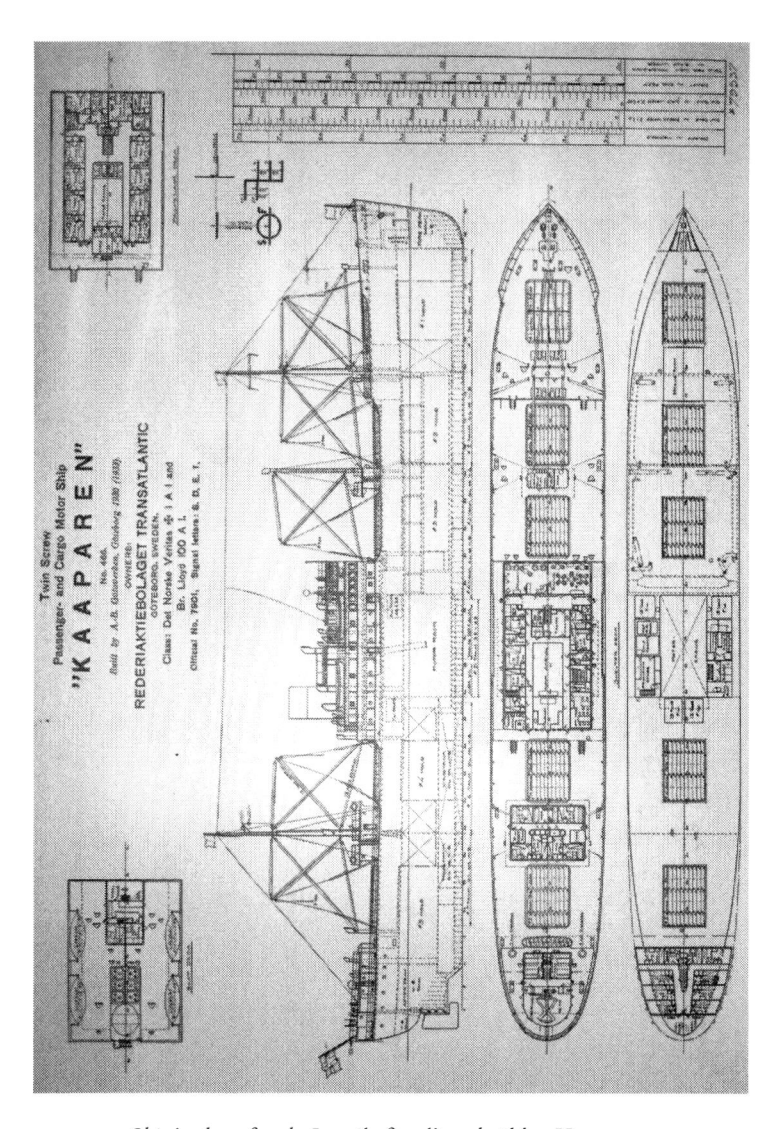

Ship's plans for the Larviksfjord's *rebuild as* Kaaparen.

Author's collection.

Diving the Kaaparen

The *Kaaparen* was rediscovered in 2000. The first divers to visit the wreck were a film crew led by Mike Fletcher of The Sea Hunters fame as well as local divers Tye Zinck, Dennis Pitts, and Paul Grantham. The wreck was extensively filmed for the television series and it wasn't long before local technical divers decided to head to the wreck.

The *Kaaparen's* wreckage sits on a sloping rock and gravel bottom at depths of 220 to 250 feet, with her bow being at the shallow end and the starboard side of the stern at the deepest point. At these depths, bottom times are short, decompression obligations long, and the margin for diver error, both in the planning and the execution of the dive, is exceedingly thin.

On Saturday, June 12th, 2004 at 8:00 am I backed my truck into the driveway of Paul Crocker's home in Dartmouth. Jason Kennedy and Mike Grebler, who were going along as our surface and in-water support team that day were already there. We quickly loaded Paul's gear into the back of my truck along with mine and jumped in the cab. We pulled out with Mike and Jason following behind and headed to Skipper Dave's Charters in Sambro, where we would be joining a group of divers heading out that morning to the *Kaaparen.*

Paul and I were both excited on this the day of what would be our deepest dive up to this point and talked during the drive of what we hoped to see and do on the wreck as well as discussing for one final time how much bottom time we'd like to do and how much decompression, that necessary evil of deep diving, we'd have to do.

Upon arrival at Skipper Dave's wharf, we found seven divers with their gear piled up on the dock waiting to be loaded aboard the dive boat *Ryan & Erin.* I backed the truck onto the dock and our gear was soon added to the pile. Loading then commenced after which, at just after 9:00, we let go the lines and headed for the *Kaaparen,* about twelve miles away.

On the boat ride to the wreck site the divers all assembled their gear, checked that everything was working properly, and then sat around on deck in the sunshine and talked about the dive to come. The group would be diving in three teams of two and a team of three, this last consisting of two rebreather divers, Nick Seemel and Paul Grantham, along with Rene Saulnier on open circuit. Mike was designated as surface support, logging the diver's times in and out of the water as well as giving Skipper Dave a hand with anchoring the boat, deck work, and passing equipment to divers after they entered the water. Jason would be in-water support, carrying extra decompression gases in case they were needed and meeting each diver at 100 feet as they ascended from their dive and taking care of any problems that may have arisen. As in-water support diver, he was also expected to be gopher and pack mule to the ascending divers. Any equipment not required during the divers decompression, from reels to lights to used decompression bottles, were clipped off on Jason's D-rings to be taken to the boat on the surface. With frequent trips to the surface he was also able to keep the surface crew apprised of the diver's status and could be summoned by the boat crew if a diver surfaced away from the boat or a surface marker popped up.

After arriving at the wreck site and hooking the wreck, the divers quickly geared up for the dive. Albert Coffill and Steve Crain were the first team to splash, and after being handed their Gavin scooters by Dave, descended the anchor line to the wreck. The team of three were the next to go, and after some initial problems with Nick's dry gloves, were also soon lost from view. Paul and I were the third team in.

Paul and I were breathing 15/50 trimix on this dive, meaning our breathing gas was a mixture of 15% oxygen, 50% helium, and 35% nitrogen. Because this was a hypoxic mix and could lead to dizziness or even loss of consciousness if breathed on the surface, especially if exerting yourself swimming against a strong current, we elected to descend the first 30 feet while breathing from our deep decompression gas, a 40% O2 nitrox mixture. At 30 feet we switched to our back gas, did a quick equipment check, and continued our descent.

Passing the 180 foot mark, we could make out the shape of the wreckage beneath us. Dave had dropped his anchor within a few feet of the massive port side diesel engine, between it and the port side of the hull. We couldn't have asked for a better land mark to relocate the anchor line at the end of our dive.

We leveled off at 225 feet and began our exploration by heading aft. We were soon past the engines and heading toward the stern of the wreck. We knew we didn't have much time and wanted to orient ourselves with the general layout of the wreck to give us a baseline for conducting later explorations. What struck us both on this quick swim to the stern of the wreck was how much the *Kaaparen* reminded us of the *Kolkhosnik*. Risdon Beazley's signature on the wreck was very apparent, with the distinctive explosive cuts along the turn of the bilge and the scattered and twisted wreckage left in the wake of the eight-legged grab. Nearing the stern we came upon the *Kaaparen's* two spare bronze propellers, each one massive at ten and a half feet in diameter. There were also many scattered ingots of aluminum, part of the cargo left behind and now glowing bright white in the near darkness and crumbling from oxidation. We reached the stern with its two propellers, the iron twins of the spares we had already passed. We turned and headed back towards the fore part of the ship, passing more discarded cargo including rubber tank tracks wound into ovals and strapped to aid in stowing them.

Once again passing to port of the engines, we headed towards the forward cargo area, passing the refrigeration room with its line of huge compressors. The forward portion of the ship was a maze of scattered wreckage and cargo. We swam over a huge mound of empty 50 mm brass shell casings, easily measuring 20 feet in diameter and 8–10 feet deep. There were thousands of them. Also in this area were huge spun-brass ingots, measuring 12 inches in diameter and about six feet in length, and more of the cargo of aluminum ingots glowing eerily with their oxidation.

There was much more of the vessel's valuable metal cargo left behind on this wreck when compared to the *Kolkhosnik*. The reason for it is simple: As mentioned in the chapter about the *Kolkhosnik*, Risdon Beazley's salvage methods routinely left about fifteen percent of the cargo behind. This was due to the fact that keeping a salvage vessel at sea is a costly venture and there came a time during the salvage where the daily rate of recovery dried up to a trickle due to most of the cargo already having been reclaimed. The little bit left and the slow rate of recovering it was not enough to warrant keeping the vessel on site for a longer period of time and so it was abandoned. On a wreck like the *Kaaparen*, which lay in over 200 feet of water, this cargo would remain. The *Kolkhosnik* however, was shallow enough for divers to work and hand-pick the remaining cargo. This means that the remaining fifteen percent of the *Kolkhosnik's* cargo could (and was) easily salvaged, while the *Kaaparen's* remained undisturbed.

As we crossed to the starboard side of the wreck, we were again reminded of the *Kolkhosnik*. Here again, was a feature left by Risdon Beazley's salvagers. Stretching away from the ship's starboard side, for farther than we could see in the 80 foot visibility, was a massive debris field, the dumping ground for those parts of the ship and cargo which were brought up by the salvagers grab but not of any value to them. Here were to be seen huge piles of discarded cargo, as well as hull plates and fittings of the ship. We noticed several of the large, rectangular brass windows from the ship's accommodation section.

We had little time to explore the debris field however, as our bottom time was rapidly being used up. As with most of my wreck dives, this one was over much sooner than I would have liked and it was time for us to head back. We crossed the wreck once again, swimming along the forward ends of the huge diesel engines and returned to the anchor line to begin our ascent.

Our first decompression stop was at 160 feet. It seemed odd to be decompressing at this depth, deeper than the recreational diving limit and much deeper than many divers would ever go. We could look down and still make out features of the wreck in the gloom below us, and every once in awhile we would see the flicker of a diver's light, like a firefly in the night.

At 100 feet we were met by Jason. We acknowledged his okay signals and assured him there was nothing we required. At 90 feet we made our first gas switch to our 40% deco mix and continued our ascent in 10 foot increments until our final switch to 80% at 30 feet. We handed off our 40% deco tanks to Jason who surfaced with them and settled down to complete the final half-hour of our decompression, now surrounded by most of the other divers. By the time we were conducting our 20 foot stop one diver, Dwayne McLaughlin, was conspicuously absent and I was beginning to feel uneasy, wondering where he was. Jason had surfaced a few minutes before and hadn't returned. We then heard the boat's engine start, the anchor line was dropped with a buoy on it and the boat took off. We continued our deco, wondering what was happening and hoping all was well. Several anxious minutes later and the boat returned, Jason splashed in and descended to where we were resting. He handed me a slate on which was written "Dwayne decoing under liftbag." I immediately felt easier in my mind. Dwayne had gone off on his own, nothing remarkable in itself, as many of us routinely conducted solo dives, but he found towards the end of his dive that he wouldn't make it back to the anchor in time. He elected to shoot a liftbag and decompress by himself under it, the bag itself being a beacon to the topside crew that a diver was there. Dave quickly collected Jason, steamed over to the liftbag on the surface and sent Jason down to check on Dwayne's status. After verifying that everything was alright, Dave again collected our by now overworked support diver, came back to the anchor line, and Jason descended to let us know all was well and put an end to our worry about Dwayne's welfare.

After surfacing and removing our gear, changing out of our drysuits, and ravenously attacking our lunch, Paul and I sat and talked about our dive. We discussed where we'd like to further explore, what areas were the most interesting and what we thought of the dive in general. I think Paul summed it up pretty well when he announced to the entire group that he'd found a new favorite wreck.

Three weeks later, I found myself back on the *Kaaparen,* this time descending to the wreck with Dwayne McLaughlin. Although we traveled down the anchor line together, we both had planned solo dives and when we reached the wreck, Dwayne headed aft of the engines while I traveled forward. We had agreed to meet back at the anchor at the end of our 25 minute bottom time and ascend to complete our deco together.

Dave had again dropped the anchor near the port diesel, only about 5 feet from where we'd been anchored the previous time. I had decided to explore only a small area of the wreck between the engines and the area of the number 2 cargo hold. By concentrating on a small area, I could take my time and search for those items of ship's fittings which were of interest to wreck divers. Items such as the vessel's telegraph, binnacle, engine room gauges, portholes, and deck fittings.

Although I scrutinized the area from the engines to about halfway to the bow, I had come up empty handed. I had purposely stayed out of the starboard side dumping area, knowing that this would require several dives to properly search and landmark. The 22 minute mark of my dive found me on the starboard side of the wreck and quite a piece forward of the engines. I took a compass bearing and headed back. The visibility was nowhere near as good as it had been on our first dive to the wreck, only about 15–20 feet, making it much easier to get disoriented and turned around in the expanse of the shipwreck. I kept a close watch on my compass and soon could see the glow of Dwayne's dive light as I neared the front of the port engine.

As I ascended to my first decompression stop at 160 feet, I watched the wreck recede into the gloom. I was not to see the *Kaaparen* again for two years.

In 2006, Paul Crocker, Jason Kennedy, and I were asked to join Greg Mossfeldt's Halifax: 2006 Expedition. The team had previously been to the area to dive some of the local deep wrecks in 2002 and again in 2003, the 2003 expedition concluding with part of the team being the first divers to visit the tanker *Athelviking*, which they videoed, sitting upright and intact on the bottom in 300 feet of water.

In July of 2006, Greg was back with a (mostly) new team of explorers, made up of divers from Alberta (Greg, Brian Nadwidny, Bryan Eaton, Jeff Grimson, and John McCuaig), Ontario (Steve Shultz and Doug Smeaton), and Nova Scotia (the above mentioned plus Harvey Morash).

On July 15th, Jason, Paul, and I arrived at Skipper Dave's wharf to find the others already in attendance and their gear already loaded on the boat. After we were introduced to those divers we hadn't yet met, our gear was added to the piles already on the boat and we cast off for our first dive of the expedition, a dive to the *Kolkhosnik*. On the following three days we conducted dives on the *L-26*, and the *British Freedom*.

Day five of the expedition found us anchored above the wreck of the *Kaaparen*. We had been blessed with good weather every day so far and this one was no exception. The ocean was like a pane of glass and the sun was shining down on us.

Paul and I were happy to be back at the *Kaaparen* after two years spent diving other wrecks. For Jason though, it was an even more memorable day: this was to be his first dive on the wreck instead of acting as support diver for the others. That role would instead be filled on this trip by Jeff Grimson. Jeff performed admirably in this difficult role, not only during this day's dive to the *Kaaparen*, but also later in the expedition on the *Athelviking* dives.

Brian Nadwidny and Bryan Eaton were the first team in the water, followed by Paul, Jason, and I. We descended the anchor line to the wreck, finding it in the sand beside the port bow. I dragged the anchor over and hooked it in a piece of wreckage and Paul tied it in. We could see Brian's bright video lights off in the distance towards the engines. We decided to stick to the forward part of the wreck for our dive. Jason led us across the wreck where we began exploring part of Risdon Beazley's dumping ground. We then headed for the very bow of the wreck, I once again marveling at the amount and types of cargo items scattered and strewn about the remains of the ship. Being Jason's first dive to the wreck, we had decided on a bottom time of 20 minutes and all too soon it was time for us to leave.

As we swam towards the anchor line I noticed what appeared to be the remains of a rotted wooden crate in the sand alongside the wreck. Inside was a pile of partly decayed animal bones. I puzzled over this as I began my ascent and it wasn't until several minutes later in the midst of making my first gas switch that I had a eureka moment as I realized that this had been part of the *Kaaparen*'s cargo of frozen meat, carried in the refrigerated numbers 2 and 3 holds.

The next morning the weather was still holding, remaining calm and clear for us. It was back to the *Kaaparen* again. This time Jason wasn't in attendance as work duties required his attention. Paul and I splashed first, again tied in the anchor and conducted a very pleasant but uneventful dive, continuing our exploration of the wreck's smashed and twisted forward section.

All the divers on the expedition with the exception of Paul, Jason, and me were conducting their dives utilizing DPV's. The advantages of using scooters when exploring deep-water wrecks was immediately apparent to us after watching the other divers and speaking to them after our dives. The amount of territory that can be covered during a single dive, not to mention the ease of transporting huge double tanks and several deco bottles as opposed to having to swim that extra gear around, quickly negates the cost of adding a scooter to your regular dive kit. With bottom time at a premium on deep, expensive, mixed-gas dives such as these, the substantial expense of a DPV will soon seem to be a small price to pay.

The following day, July 21st, was to be my last day with the team, other commitments precluding my participation in the remainder of the two-week expedition. The dive that day was to be to the *Athelviking*, with both Paul and I looking forward to conducting our first dive on that wreck. But it was not to be. The weather gods had graced us with six straight days of perfect diving conditions and now it was time for payback. We were blown out the next day and we said goodbye to the rest of the team for that season. As it turned out, the bad weather continued for 5 days and it wasn't until July 26th, with only two days left in the expedition, that the team finally made it to the wreck of the *Athelviking*. But we had gotten in six fantastic days of diving (eight for the rest of the team) with two of those days on the unforgettable wreck of the *Kaaparen*.

The Team 2006 divers on the Ryan & Erin *after the first* Kaaparen *dive.* **L to R: Standing**: *Bryan Eaton, Paul Crocker, Jason Kennedy (in front of Paul), Al Henneberry, Brian Nadwidny, Greg Mossfeldt, Jeff Grimson.* **Seated**: *Steve Gray (*Ryan & Erin *crew), Harvey Morash, John McCuaig, Steve Shultz, Doug Smeaton.*

David Gray photo.

Chapter 8

The "Mystery Schooner"— Hebridean?

Diving the "Mystery Schooner"

On August 27th, 2004, five other divers and I were aboard Skipper Dave's dive vessel *Ryan & Erin,* heading on this beautiful late summer evening to a position in the Approaches to Halifax Harbour some distance east of Chebucto Head, near where the pilot boats meet inbound ships.

We were going to dive a wreck that had been dubbed the *Mystery Schooner* by Skipper Dave Gray. The wreck had been discovered only a few years previously and had been little visited since. It had yet to be identified. Excitement was rising among the group as we neared the site, each diver deciding he would be the one to discover the clue that would lead to the positive identification of the vessel.

Dave slowed the boat as we neared the wreck site. As Paul Crocker and I completed gearing up, Dave dropped the hook. We waited a few minutes for the boat to take up the slack in the anchor line, and the anchor began to drag. Dave hauled the anchor back in and made another pass over the wreck, once again dropping the anchor into the wreckage. This time when the anchor rode came tight, the hook held. Paul and I wasted no time rolling over the rail and descending the polypropylene line to the wreck.

As we neared the wreck in the deep-water gloom created by the nearly setting, early evening sun, we began to make out shapes ... first the main engine, then part of the wreck's stern, the rudder, the prop.

Dropping to the bottom at 153 feet I saw that the anchor had hooked the brass tail shaft forward of the 60-inch bronze prop and was scraping and bouncing with the motion of the boat above, threatening to tear loose. I quickly took a couple wraps around the shaft with the tie-in line attached to the anchor's claws and then fastened it back to the anchor. The dive boat was now securely tethered to the wreck and we could begin our explorations.

Swimming forward we soon discovered the vessel's diesel engine, a huge monster for the size of the vessel by today's standards, and a clue as to the wreck's age. Large, slow-turning diesels such as this were used in coastal and fishing schooners from about the mid-1920's up to the 1950's. Even before diving this wreck, our friend Jason Kennedy had an idea as to its identity and the vessel's engine certainly placed the wreck in the correct time frame.

We continued forward, noting the almost complete deterioration of the hull, thus adding another clue to the admittedly tiny trail of identifying factors. A mid-20th century vessel, to be deteriorated so badly in water of this depth, meant that it was probably locally built, of native softwoods which would break down much faster than if the ship had been built of hardwood.

There were enough identifiable features however, to definitively say that this was an early mid-century two-masted schooner with a diesel auxiliary. We swam past the main mast and fore mast deadeyes, a couple still attached by their bands to the remains of the rails at the chain plates. There were also galvanized steel belaying pins in evidence, again an indicator of the wreck's age as these would be constructed of hardwood on an older vessel. Continuing forward we discovered the anchor windlass, the type conforming to those used on locally built schooners. As we passed the stem of the wreck, we began seeing what appeared to be projectiles lying on the bedrock off the wreck site. We followed what appeared to be a trail of unexploded ordinance away from the wreck for several minutes without coming to the end of the line.

Upon later discussion of these out-of-place munitions, two theories as to

A rigging deadeye and iron belaying pin recovered from the Mystery Schooner *during one of the early dives to the wreck.*

Author photo.

its proximity to the wreck emerged. The first being the simplest: That it was simply a coincidence that this ordinance from an unrelated incident came to rest alongside this wreck. The second being that the schooner had been involved in a collision and the other vessel involved had lost part of its cargo of munitions after the collision occurred. I didn't put much stock in the second theory because any steel-hulled vessel colliding with a small schooner would likely have received little damage from the encounter. For the time being it would remain a mystery as our bottom time was quickly being used up and Paul and I turned and headed for the wreck again and the waiting anchor rode to begin our decompression.

We reached the wreck again at its port side near the bow and retraced our fin strokes toward the stern. As we were passing the engine, I noticed a large brass fire extinguisher under the engine on the starboard side. I extricated the crushed artifact and carefully scanned it for any markings or date. There was nothing, not so much as the name of the company who had manufactured it. We weren't going to identify the wreck this way. Another of these fire extinguishers had been recovered on a previous trip to the wreck but like this one was found to have no identifying marks. I left the artifact there and we continued to the anchor line, passing the third team of divers, Rene Saulnier and Nick Seemel, just beginning their exploration of the wreck. Jason Kennedy and Nigel Leggett had been the second team to reach the schooner and were sifting the wreckage for clues as Paul and I began our ascent to our first decompression stop at 90 feet, from where we could watch the lights of the others playing over the wreck like fireflies in a summer field.

We completed our uneventful deco, climbed to the dive boat's deck and awaited the return of the other divers, hopefully with some clue to the wreck's identity in hand. No indicator was found, the wreck was holding on to its secrets.

The small wreck had been found by Mike Caudle in 2000. Mike had been conducting a magnetometer search in the approaches when he obtained a hit from the engine of this wreck. Lowering a drop video camera to the wreck, deadeyes were seen on the video monitor. That first night, Mike and Tye Zinck geared up and conducted the first dive on the wreck, stating that it was in fact a sailing vessel, but not an extremely old one as evidenced by the diesel engine sitting squarely in the wreckage.

Subsequent dives by Mike, Tye, and others yielded artifacts including a ship's bell.... .but none of these contained markings to identify the ship and it was left alone in the pursuit of other wrecks, namely the *Kaaparen* and *Clayoquot*. And so the little wooden wreck became known as the *Mystery Schooner*, although there were clues as to its identity.

That night in August of 2004 as we returned to shore, a visitor was waiting for us on the dock: Mike Caudle had heard we were out trying to identify "his" wreck and was curious as to whether we'd had any luck. We replied in the negative, but began talking of what it could possibly be while unloading our gear from the boat. Jason then mentioned his theory to Mike, that he believed the wreck to be the ill-fated *Hebridean,* which Mike admitted was precisely what his research had led him to believe.

Unfortunately, even after several additional dives to the wreck, conclusive evidence as to the identity of the Mystery Schooner remains elusive. The circumstantial evidence however, has pointed like a compass needle directly towards the *Hebridean*. Jason Kennedy, who has done much of the research on the *Hebridean/Mystery Schooner,* and I believe that we've identified the wreck, but we continue to search for the concrete proof that will allow us to definitively say, "Yes, it is the *Hebridean."*

History

The schooner *Hebridean*, official number 154615, had been built in 1928 in Mahone Bay, Nova Scotia. She was 92 feet in length by 23 feet wide and 98 gross registered tons. The schooner was two masted as were most of the coastal schooners at the time and was fitted with an auxiliary diesel engine.

The *Hebridean* was also known as Pilot Boat #2 for she wasn't a fishing schooner as were most of her sisters, but instead carried the pilots who guided foreign ships safely into Halifax Harbour.

By March of 1940 the vessel traffic entering and departing the harbour had grown to staggering proportions, with huge convoys departing for England carrying war materials and convoys of empty ships returning to have their bellies filled once again to continue feeding the ever-hungry Allied war machine. Convoys of as many as sixty ships were sailing out of Halifax under escort every 4–5 days.

On the evening of March 28th, 1940, just after 9:00, the *Hebridean* left her home port of Herring Cove with nine pilots and six crew aboard. They would spend most of the night east of Chebucto Head waiting for arriving vessels and putting an experienced pilot aboard each. As the night progressed, pilots and crew amused themselves while they waited, reading, yarning, playing cards, or listening to the radio in the cabin.

Eventually, the lights of a ship were sighted. As the vessel drew nearer, the crew of the *Hebridean* prepared and launched a dory which would take one of the pilots to the ship when it reached the pilot station and slowed to allow the man to board.

Shortly after midnight, pilot Tupper Hayes, and crewmen Edward Laughlin and Walter Power boarded the dory and stood off to wait for the vessel to slow enough for Tupper to board it.

The incoming vessel was the *SS Esmond*. She had been built for the Law Shipping Company in 1929 and christened the *Traprain Law*. In 1932 the ship was transferred to the Anglo-Newfoundland Steamship Company and renamed. By that early morning of March 29th, the 4976 ton *Esmond* had already made three round trip crossings of the Atlantic since the war began, and was hurrying to join her fourth Halifax-England convoy, HX 31, which was to depart that day. The ship had just made the voyage alone down the coast from Newfoundland where she had taken on a full cargo of newsprint.

As the three men in the dory looked on, the *Esmond*, rather than slowing as she neared the pilot station, kept steaming ahead at near her best speed. With horror they realized that the big steamer would collide with the *Hebridean*, so lately vacated by them. Any evasive action that may have been taken by the *Hebridean's* crew came too late, and the little schooner was run down where she sat.

Above: *The* Hebridean *at dock in Lunenburg.* Photo courtesy Hubert Hall, Shipsearch Marine. **Below:** *The vessel which ran down the* Hebridean, *the* S.S. Esmond.

Author's collection.

The *Hebridean* sank almost immediately, taking with her the bodies of six pilots and three of her crew. All of the men lost in the tragedy were from the Halifax and Herring Cove areas. Three men in the water and the three in the dory were picked up and landed at Halifax.

The Duncan's Cove lifeboat was sent out along with a tug from the naval dockyard in Halifax. Although they searched the area surrounding the sinking for hours on end, no trace of the other men was found.

The *Esmond* made her ignominious way into Halifax Harbour and to her anchorage in Bedford Basin. She would miss her convoy that day and would instead sail as part of Convoy HX 32 on April 2nd. *Esmond* herself wouldn't live to see the end of the war. On May 9th, 1941 the ship was part of convoy OB 318 and was steaming to the south of Iceland when she was torpedoed by the *U-110*. Like the *Hebridean* the previous year, the *Esmond* went to the bottom, but unlike the little schooner, all forty-five of her crew and five naval gunners survived.

Two headstones in a Herring Cove cemetery give mute testimony to the families that were changed forever by the tragic sinking of the *Hebridean*. Nine women were widowed that night and thirty-five children lost their fathers. The inscriptions on the two headstones read:

Eva Francis Pelham
1896–1986
Wife of Lionel Francis
Lost on "Hebridean", 1940

And:

Elsie Agnes Dempsey
1899–1984
Husband Carl F.
Lost in "Hebridean"
1900–1940

The nine men who lost their lives that cold March night were James Renner, James Dempsey, Lionel Pelham, and Claude Martin of Halifax. Loran Dempsey, Carleton Dempsey, Matthew Power, and Lawrence Thomas of Herring Cove, and Roy Purcell of Portuguese Cove.

Return to the "Mystery Schooner"

In 2006, we decided it was time to identify the wreck. Our first dive that year to the wreck we believed to be the *Hebridean* was conducted on June 25th. The wreck diving season was already well advanced and we left the *Ryan & Erin's* dock determined to solve the mystery and positively identify the wreck.

This time there were only four divers and we took care of our pre-dive preparations as Dave was anchoring the dive boat over the wreck. It was decided that Jason Kennedy and Nigel Leggett would splash first and do the tie-in. Over the side they went while Paul and I finished gearing up. Before we could splash in, the first team had returned to the surface: the anchor had broken free of the wreck and the boat was dragging it across the bottom with the wreck nowhere in sight. Dave hauled the useless hook back in and made another pass.

Once again the hook was dropped and all four divers splashed almost together. We quickly made our way to the bottom and found the anchor hooked near the stern of the wreck. Jason quickly tied in before he and Nigel headed towards the bow of the wreck in the early evening gloom of 150 feet of water. Paul and I headed in the other direction with a different plan in mind.

I tied off my wreck reel on the schooner's propeller and started running out the line as we swam away from the wreck with the intent of making a 270 degree sweep off the wreck's stern. Our plan was twofold for this dive: First, we wanted to ascertain whether there were any dislocated pieces of wreckage or artifacts away from the wreck which might help to identify her. Second, we wished to see if the ordinance noticed previously extended off in this direction as well and whether the trail petered out or not.

We ran out over 100 feet of line and began our sweep from the port side, around the stern and to the starboard side of the wreck with me at the end of the line and Paul about 35–40 feet closer in to the wreck so we could visually cover the entire area. We saw nothing that could aid us in our quest to put a name to the wreck. We did discover however that the projectiles were scattered around the bottom in this direction as well. This information confirmed that the ordinance was scattered around the wreck in at least three directions.

As there appeared to be nothing visible during our search to help us identify the ship, I began to reel in my line and we headed back to the wreck to see how the search was going for the other team. As I stowed my reel we began swimming forward along the port side of the wreck but out from it about 15–20 feet. Jason and Nigel were closely inspecting the wreck along the starboard side. Paul and I swam to the anchor windlass when he indicated he was heading for the upline. I signaled that I was going to stay another five minutes or so and bade farewell to my dive buddy as he turned and retraced his path along the schooner's length.

I continued to the wreck's bow, looking closely into holes, cracks, and crevices for any artifact or clue as to the wreck's identity. By the time I was thirty-two minutes into my dive I had decided that it was time to go, it was obvious that the wreck wasn't going to tell me her secret on this dive. I reluctantly headed back to the dive boat's anchor line, our highway to the world of light and warmth, watching small pollack and cunners flicker through the beam of my dive light in the gathering dark. Hopefully one of the others would have had better luck than Paul and I. I left the bottom at 41 minutes run time, heading for my first deep stop at 125 feet, the last to leave the wreck that night.

After completing my decompression and climbing the ladder to the *Ryan & Erin's* deck, I sat heavily on the bench where eager hands helped me doff my deco tanks and doubles. After answering in the negative as to my success in solving the wreck's identity crisis, I was answered by the others in a similar fashion: once again the *Mystery Schooner* had stymied us.

With Greg Mossfeldt's 2006 Expedition looming in mid-July it would be almost two months before we'd return to the wreck. Later in August and September we completed three more dives on the *Mystery Schooner*. It still refuses to tell us its name. Although Jason Kennedy and I are certain the wreck is that of the doomed *Hebridean,* we will continue searching the remains of the ship for the proof we seek until we're finally successful and can put the mystery to rest once and for all. My hope for now is to keep hunting through the wreckage and be able to write in a future volume that we have given the *Mystery Schooner* a real name.

Chapter 9

Disaster in Chedabucto Bay: S.S. Arrow

History

The oil tanker *Olympic Games,* official number 260538, slid down the ways and met the waters of the Atlantic on August 27th, 1948 at Bethlehem Shipbuilding, Sparrow's Point, Maryland. The ship was designated as hull number 4463 by the shipyard and was completed and turned over to her owners in November of the same year.

The tanker was 529 feet in length by 63 feet in breadth and 11,379 gross tons. She was powered by a single steam turbine engine driving a single propeller.

The *Olympic Games* was commissioned and owned by Sunstone Marine Panama Ltd. and managed by Olympic Maritime. The tanker was one of sixty-some tankers managed by Olympic, with each one belonging to a separate company, and all of the companies allegedly owned by the Greek tycoon Aristotle Onassis. Collectively these ships were known as the Onassis fleet even though on paper each had a different owner.

In 1960 the ship was renamed *Sea Robin,* and renamed once again two years later, this time becoming the *Arrow.*

The *Arrow,* at twenty-two years of age, was the second-oldest tanker in the Onassis fleet when she left Amuay Bay in Venezuela early in 1970. She had just finished loading 16,000 tons of Bunker-C oil and was making the long trip north to deliver her cargo to the Nova Scotia Pulp & Paper Mill at Point Tupper, near the Canso Causeway on the Cape Breton side of the Canso Strait.

Captain George Anastassopoulos and all but one of his crew of thirty-four men were from Greece, although most of the officers had spent considerable time in Canadian waters and knew the coast fairly well. The captain himself had only taken command of the *Arrow* on December 11th of 1969, having joined the ship in Charlottetown, Prince Edward Island just before she left on the journey to Venezuela. Upon taking command he found a vessel ill-equipped for the voyage ahead and with many necessary repairs having been continually put off in the name of profit margins.

Over the course of the voyage the tanker's gyro compass failed and her radar was working only sporadically, making it almost completely unreliable. She carried no Decca Navigator or Loran A electronic navigation systems (standard bridge equipment on ocean-going vessels at the time) and her emergency generator was unserviceable. Her echo sounder as well, had not been working since her captain had taken command in December. There were also no Canadian charts aboard for the area she would be sailing, the treacherous region of Chedabucto Bay and the approaches to the Canso Straits. Without the gyro compass the ship was being steered by her magnetic compass. Since this had not been recently adjusted there was no way of knowing how much magnetic deviation was present or how much the total compass error was (compass error equals magnetic variation +/- deviation). The combination of an unreliable radar, a complete lack of on-board electronic navigation instruments, and an unknown magnetic compass error had made the tanker's navigation nothing more than a series of hopeful guesses that didn't even amount to proper dead-reckoning. As she approached the rocky coast of eastern Nova Scotia the ship was navigationally little better off than Columbus' fleet had been during its historic voyage 478 years before.

The *Arrow* entered Chedabucto Bay early on the morning of February 4th, 1970. The southerly wind was blowing 35–40 knots and kicking up 10–12 foot seas as the ship headed up the Bay towards the Strait of Canso and Point Tupper. Captain Anastassopoulos and one of his officers kept a lookout for the treacherous Cerberus Rock for after passing this hazard on their starboard side they could relax and enjoy the final approach up the Strait to Point Tupper. (Cerberus Rock was named after the monstrous three-headed dog that guarded the gate to Hades in Greek mythology.)

The prevailing wind and sea conditions were the final factors which combined with the *Arrow's* deficiencies into what would become Canada's first and to date worst environmental disaster. The tanker was steaming on an approach of 312 degrees by her magnetic compass which should have kept her safely south of Cerberus Rock. After safely passing this hazard, she'd have altered course to approximately 340 degrees magnetic for the run up the Strait. Had her gyro compass been working there would have been no problem as the ship would have followed true courses of 290 degrees and then altered to 320. (For those not familiar with ships and their navigation, gyro compasses point to true north unlike magnetic compasses which point to magnetic north. The difference between the two are made up by variation [error caused by the Earth's magnetic field] and deviation [error caused by the ships' magnetic field]).

The problem was that the magnetic deviation of the *Arrow's* compass was not known due to the lack of any regular maintenance aboard the vessel. The last time the compass had been adjusted was not known and the deviation card for it was woefully out of date. While the vessel was merrily steaming along on her supposed course of 312 degrees she was actually making good a course of approximately 320 degrees and steaming directly for Cerberus Rock, unseen by the lookouts on the bridge in the inclement weather, and which she struck at a speed of twelve knots at just past 9:30 Atlantic Standard Time on February 4th, 1970. Unfortunately, the grasp applied by the mythical dog on the ships' bow was much firmer than the captain realized.

Shortly after the ship struck a call was placed to the tanker's agent in Port Hawkesbury requesting a tug to be sent out to assist the vessel in getting off the rock. There was no immediate severe damage and the captain believed that with the rising tide and a tug assisting, the tanker could be refloated fairly easily with little damage to her structure and with her cargo remaining aboard and not having to be lightered into barges.

At 1230 Canso Radio was finally notified and this was the first inkling Canadian Authorities had of the accident. The *Arrow's* transmission to Canso stated:

TANKER ARROW TO CANSO RADIO STOP AGROUND CERBERUS ROCK CHEDABUCTO BAY STOP NO IMMEDIATE ASSISTANCE REQUIRED STOP LOADED WITH FUEL OIL

Later that afternoon Canso Radio overheard a radiotelephone conversation between the *Arrow* and the pilot boat stating that the *Arrow* was leaking oil into the sea from her starboard side. Canso contacted the *Arrow* to ask if she required assistance and was told, "Not at this time. Will call if I do."

Later that night matters were becoming dire and the crew moved to the stern of the ship as her forward section was nearly awash. By 0200 on the morning of the 5th the deck was completely awash and the ship was listing heavily to port. The crew was taken off by the Coast Guard vessel *Narwhal* at 0600 and the *Arrow* was left abandoned until a crew from Atlantic Towing led by salvage master Walter Partridge arrived later in the morning to take stock of the situation.

The Arrow *aground on Cerberus rock.*

Government of Canada photo.

What Partridge found when he boarded the ship shocked him. The captain's reports had given the impression that the damage was slight and that he would have a relatively easy task of taking the ship off the rock with her cargo in her own bottom. He was amazed to find that the tanker was in fact open to the sea and tidal in most of her cargo tanks and that oil was pouring into the bay.

Although efforts to patch the myriad holes in the tanker's bottom met with little success, one bright spot was that her engine and boiler rooms had so far remained dry. Steam was raised and the spirits of the salvage gang were lifted by the promise of being able to pump the *Arrow's* cargo into lighters and still have a decent chance of saving the ship.

By the morning of February 7th those hopes were fading fast. The *Arrow's* chief and 2nd engineers, who had been on watch throughout the night, were exhausted and the fires under her boilers were allowed to die. The *Arrow* had become a "dead ship" for the second time in three days.

All through this time the *Arrow* had been bleeding out her cargo, with thousands of gallons of heavy bunker C oil pouring into the waters of Chedabucto Bay. On the afternoon of the 8th the tanker broke in two, causing further quantities of oil to run out of her tanks. The entire section forward of the cargo masts including the forward accommodation area and bridge sank with all of the cargo tanks open to the sea.

Walter Partridge decided the stern of the tanker, which remained partially aground, could still be saved, but that it would require 600 tons of oil to be pumped from her #7 tank and be distributed among the other tanks and then pumping in compressed air to achieve the required buoyancy. This was a tried and true method of salvaging a severely holed ship in which the damaged vessel was literally floated off the rocks on a cushion of air. But salvaging the *Arrow* was not to be so easily accomplished.

On February 9th with the work progressing at a feverish pace and the salvage gang being driven to near exhaustion without a break, a weather report was handed to Walter Partridge as he was overseeing the work progressing on the *Arrow's* deck:

> *Environment Canada Marine Weather Forecast. Forchu and Chedabucto Bay. Gale warning issued. Light winds increasing to SE 20 knots this evening. Winds increasing to easterly gales 35–45 knots by Tuesday evening.*

This was a serious development. But as a professional salvor Partridge had dealt with this type of challenge before. He set to work ensuring that there were sufficient tugs to hold the *Arrow* in place through the weather of the next two days.

During the gale, the stern section came free of the rock but because of the severe weather the decision was made to hold the Arrow in the vicinity until the weather moderated rather than try to reach a safe, sheltered haven with an ungainly half-ship being towed by the stern and risking having the tanker strand herself again.

On February 12th, the captain of the Foundation Maritime tug *Foundation Valiant*, which was standing by the wreck, called in to say the *Arrow's* stern half was in trouble. She was now listed to 45 degrees and water was pouring in through hatchways and ventilators. Within minutes of that report the *Arrow* sank, coming to rest on the bottom, upright with only her funnel showing above the surface.

Before the stern sank, about half of the *Arrow's* 16,000 ton cargo of Bunker C had spilled into the ocean, devastating the coastline around Chedabucto Bay, and killing untold thousands of seabirds, marine mammals, fish, and other marine organisms as well as affecting the habitat of these animals for years to come. Also affected were the livelihoods of the fishermen and their families from the surrounding coastal communities. A full 8,000 tons of oil remained in the cargo tanks of the sunken stern section of the wreck. In order to reduce what would surely be a larger insult to the marine environment the salvors opted to undertake a historic operation: The *Arrow* disaster would mark the first time that oil cargo would be removed from a sunken tanker.

A view of the ship from forward before the bow separated. It's plain to see that her back is already broken and an oil slick can be seen in the background. The heavy bunker C oil can also be seen on the white paint at the base of the bridge superstructure.

Government of Canada photo.

In a process known as "hot-tapping", navy divers were brought in to connect hoses to the wreck's cargo tanks and the oil was then pumped into the cargo tanks of the barge *Irving Whale* which was moored above the wreck. Retrieving the oil from the sunken vessel was a Herculean task but with the salvage crews and a crew of navy divers giving their all, the mission was finally completed. The oil then made its way to the Imperial Oil refinery in Dartmouth and the *Irving Whale* herself later sank off Prince Edward Island, thus creating another oil-filled hazard in the Atlantic, the loss and recovery of which is a fascinating story in itself.

The cleanup of the *Arrow's* spilled oil took many months and cost over three-million dollars. Only about half of the 150+ miles of polluted coastline was cleaned, the rest becoming a legacy to the disaster. Even today relatively large patches of oil, now hardened into an asphalt-like substance, can still be seen around the shores of Chedabucto Bay, mute testament to the environmental calamity needlessly caused by the *Arrow's* faulty equipment and man's insatiable thirst for petroleum products.

The *Arrow* herself was left to lie beneath the waves of Chedabucto Bay, slowly deteriorating in the harsh marine environment and occasionally burping out a small quantity of the oil which remained trapped in the upper reaches of her cargo tanks and hatch coamings. Although the memory of the disaster would never be erased, the wreck itself was forgotten except for a small group of sport divers who began to visit the tanker's grave as early as the mid-1970's.

As the sport of scuba diving took a greater hold on the general public in the 1980's, the wreck became a popular dive destination, with some groups through the late 80's and even up to the present, making annual pilgrimages to the *Arrow's* grave.

Diving the Arrow

On the Labour Day weekend of 1994, a 4-day, 6-dive trip to Arichat to dive the wreck of the *Arrow* was planned by Jim Henneberry (no relation to the author), owner and operator of Dive Masters Scuba Shop in Dartmouth. Jim had visited the tanker many times, and from the late 1980's until the mid-1990's he put together an annual trip, by invitation only, to dive the wreck on the Labour Day weekend.

Pete DeGrace and I were novice divers at the time, having only been diving for a few years by this time, and felt honoured to have been asked along on the trip which was filled with divers who had been our instructors and dive masters when we had taken our courses. We were awed to be included in such lofty company.

In those days, Pete and I always dived together, and often recounted to others that we knew each other so well that each could tell what the other was thinking underwater. Although this closely-connected pair of divers had completed many wreck dives together by this point in our diving careers, we had yet to visit the *Arrow* or undertake any wreck penetration which would be possible there but wasn't an option on the storm-flattened and dynamited wrecks near Halifax. We looked forward to the trip and by Friday afternoon when Pete got off work I had our gear packed, checked, double-checked, and ready to go.

After a quick bite to eat we bid farewell to our families and took off in Pete's Honda (extremely overloaded with all our gear for three days of diving and camping) on the four-and-a-half hour drive to Arichat on Cape Breton's Isle Madame. The rest of the group had gone up earlier to get in a dive on Friday afternoon, but because of scheduling, we were forced to miss the Friday dive and would arrive Friday night to have a beer with the gang and then head out to the wreck bright-eyed and bushy-tailed on Saturday morning.

We arrived at the camp site on Isle Madame around 9:00 that night, with Metallica blaring from the car's stereo, pumped up and ready to dive the next morning. A quick call home to let everyone know we had arrived soon changed our plans however: Pete found there had been an emergency at home since we left and said he was leaving again right away. I elected to do what we had learned in our dive training and stick with my buddy. Pete and I were not only dive buddies, but also best friends; anything that affected him affected me as well. We made our quick apologies to the rest of the divers, jumped back in the car and were away on a record-breaking trip to Halifax with Pete driving like a race-car driver and me lighting us both a steady stream of cigarettes (we were both pack-and-a-half a day smokers at the time). When we arrived at Pete's place later that night, we found to our relief that all was well, but we would have to wait another year before we would see the wreck of the *Arrow*.

Exactly a year later on the Friday afternoon at the start of the Labour Day weekend we were sitting on the government wharf in Arichat waiting for our chartered dive boat to arrive, but there were crossed wires with the scheduling and our boat captain, Dave Boudreau, was out hauling crab pots. After several

Left: *The port alleyway looking forward.* ***Right:*** *Jason Kennedy and Paul Crocker swim along the port alleyway.*

Mike Grebler photos.

phone calls back and forth to Captain Dave on the boat, our dive vessel, the *Stormy Weather*, finally appeared, complete with enough snow crab left aboard after unloading her catch to feed the population of a small country (this was to be our dinner, compliments of Dave himself). It was by this time very late in the day, so we elected not to dive and instead divers and crew sat around drinking beer, eating snow crab, and spinning yarns about diving and any other imaginable subject. Even if we couldn't be underwater, this was still a pretty fine way to begin a dive trip.

Early the next morning we were tied into the stern of the wreck and getting ready to dive. Jim and his long-time dive buddy Terry Preeper would be the first to hit the water, they were heading for the wreck's engine room and would be making the longest and deepest dives (with the engine room's depth bottoming out at about 97 feet, it's 12–15 feet deeper inside the wreck than on the sand bottom outside).

Pete, Scott Maxwell, and I were the next team in. Scott had visited the wreck several times before and was giving us the "*Arrow* Grand Tour" on this, our first dive on the tanker.

We descended the buoy line to the after-deck, sitting level at only 35 feet. The deck was a mass of equipment, with winches, blocks, fairleads, and mooring bitts all in evidence as we dropped over the stern for a look at the massive rudder. The propeller shaft itself ended in a jagged stump, the prop having been salvaged shortly after the tanker sank. We dropped to the sandy bottom at 85 feet, looking up at the huge shape of the wreck towering above us. After swimming over the wrecking-yard piles of debris that were the shallow wrecks near Halifax, seeing this monstrosity standing up from the ocean bottom would have rendered Pete and I speechless, had that not already been taken care of by the surrounding sea and our regulator mouthpieces. We swam to starboard of the stern and back up to deck level, reaching the afterdeck and heading forward to the remains of the after accommodation. Although much of the aft superstructure was still standing at this time, twenty-five years of harsh North-Atlantic storms had taken their toll and although individual cabins as well as the galley area were in evidence, the bulkheads and decks were riddled with holes as the old tanker quietly rotted away in her watery grave.

We followed Scott through the galley and to the break of the aftercastle, where a ladder led down to the after tank deck. Swimming through a doorway I found myself in the after pump room, with its dizzying array of valves and levers. I marveled at how some member of the crew had been responsible for this area, and knew the function of each valve, lever, and switch as part of the routine of his job. Pete followed me in, had a quick look around, shook his head at the sheer number of valve handwheels and ducked back out. I followed him out and found Scott patiently waiting to continue forward.

We resumed our leisurely exploration, swimming along the tank deck past huge valve handwheels, each marked with a copper tag stating its function. We followed the transfer piping along the deck until that point where we should soon arrive at the midship's superstructure, which of course we knew was missing, and came to the jagged break in the deck where the stern had broken from the bow. We swam out past the break and slowly freefell to the bottom, turning around to gaze into the huge gaping maw of the now empty cargo tanks where the entire ship's structure had sheared off as if made of paper instead of heavy steel. At the bottom we found scattered hull and deck plating where the wreck was slowly collapsing at its forward end.

Left: *The stern of the* Arrow *rises above the flat sand and gravel bottom of Chedabucto Bay.*
Right: *Janice McDougall hovers over the wreck's massive rudder.*

Mike Grebler photos.

Janice McDougall swims through the remains of the tanker's galley. This shot shows the area of the port alleyway two years after the photos on page 90. The bulkhead is now gone allowing a straight swim into the galley area from the alleyway.

Mike Grebler photo.

    We ascended once again to the main deck and began our swim aft towards the upline. When we reached the port ladder leading up to the aftercastle, Scott removed his fins and climbed the ladder, pausing at the top to replace his fins before following Pete and I through the port side alleyway and back to the afterdeck. We arrived at the deck winch to which the line was tied and made our way to the surface. In our first 80 minute dive to the wreck we had seen the entire length of the *Arrow's* stern section including the unforgettable sight of staring up at the upright ship from the bottom of Chedabucto Bay. We couldn't have been happier with the dive and could barely wait out our surface interval to get back and explore the crew's accommodation under the after deck.

On our second dive, Pete and I headed for the lower deck accommodation area, beneath the main deck aft. There were two deck levels here with cabins for the crew and several storage rooms. We tied off a reel at the break of the aftercastle and swam aft, running out line as we went, ducking into cabins and washrooms and gazing out of the now glassless portholes. After exploring as far aft as possible, we retraced our path and exited the wreck. After so many divers combing the wreck for artifacts over the years, these were becoming scarce, but we did manage to find a bronze cage light and two brass door knob sets on this dive, souvenirs of our visit to the wreck.

We would make two more dives to the Arrow the next day before heading home while the rest of the team stayed to get in one more dive on Monday morning. On our second day we each recovered a porthole from the wreck to add to our list of artifacts. We had a fantastic weekend of diving but I would have to wait five years before seeing the Arrow again.

There was no *Arrow* trip in 1996. For both the 1997 and 1998 trips I had other commitments and could not make the pilgrimage to the wreck. Finally, in the summer of 2000 I made it back to the *Arrow* again for two days of diving.

On the first day we were blown out so our four dives had just been cut to two. On our first dive the next morning, Paul Crocker and I hit the water first and descended to the wreck. The change in the *Arrow* over the five intervening years was startling. Most of the aft superstructure was now missing, or so riddled with holes that it resembled a block of Swiss cheese. The deck over the engine room had partially collapsed down upon the engine and the huge cavity this created yawned like the mouth of some rusty-toothed sea beast.

Paul and I picked our way through the hanging deck plates and began a tentative penetration of the engine room. Besides the darkness, the inclement weather of the day before had reduced visibility and inside the wreck the sediment had not yet settled. Our powerful dive lights were nearly useless as we crept through the room mostly by feel. We were looking for a storage room in the aft part of the engine room but it was like trying to find your way through a maze at night while wearing a blindfold. At one point I felt a doorway and thought, "Here's the store room."

This is the galley area looking aft (a deck winch on the after deck can be seen through the doorway). The many gaping holes in the bulkheads and deckhead show just how quickly the wreck is now deteriorating.

Mike Grebler photo.

I swam through and found myself in a room no bigger than a closet, where I had not even enough room to turn around. And Paul was trying to push his way in behind me. Somehow in the nil visibility I communicated to him that this wasn't what we were looking for and he backed out, allowing me to exit. We continued on but after a few more minutes decided we were wasting our time and followed our line back to the exit.

We then took a swim out along the tank deck, a swim which was much shorter than I remembered it. Over the intervening five years since I had last visited the wreck, much of the cargo deck had collapsed and the intact stern section of the *Arrow* was now at least 100 feet shorter than it had been in 1995. Such is the destructive power of the sea, eventually turning even the most intact, pristine shipwreck into nothing more than a scrap yard pile of junk.

We swam back along the deck to our upline and ascended to the surface where the trusty *Stormy Weather* was waiting for us.

On our second dive that day we circumnavigated the wreck at 80 feet, noting the damage done by the ever-hungry ocean over the past few years and near the forward part finding ourselves swimming over a twisted mass of collapsed deck and hull plating that seemed to go on forever.

Diver Georg Knoepfler contemplates the piles of dive gear and tanks on the deck of the Stormy Weather *during the 2000 trip to the* Arrow.

Mike Grebler photo.

Since then, the wreck has continued to deteriorate at an ever increasing pace, the tired old steel finally wearing out after 37 years immersed in the corrosive bath of the sea. While the stern and part of the tank deck are still holding together for the time being, it won't be much longer before the Arrow becomes another of those shallow-water wrecks which are only flattened piles of rotting hull plates. I feel fortunate to have been one of those who could dive the wreck in the days when it remained mostly intact and penetration of the engine room and crew's quarters was still possible in relative safety.

Chapter 10

Empire Kingfisher: Risdon Beazley's First Canadian Salvage Operation

Diving the Empire Kingfisher

On a Sunday morning in early February of 2007, Sam Millett and I sat in a Tim Horton's on Wyse Road in Dartmouth sipping coffee and talking about, what else?, diving. We had planned a bottle dive in Halifax Harbour that morning, but with severe weather conditions and biting cold we decided that discretion was indeed the better part of valour (or at least of sanity) and opted for what we called a "Tim Horton's dive" instead, sitting in a warm coffee shop telling diving stories.

Our discussion on this particular morning had turned to the notion of a possible dive trip to St. Paul's Island off the northern coast of Cape Breton later that summer. I had for some time wanted to put together a trip to this remote and little dived area.

While we were thus engrossed, our long-time dive buddy Jason Kennedy (who had wisely bowed out of even thinking about diving that morning) joined us for coffee. The three of us kicked around the idea of a St. Paul's trip. The island is known as the "Graveyard of the Gulf" and we knew that a trip there would give us plenty to choose from in the way of wrecks to dive. Sam had been involved in several expeditions to the island and told us from experience that there were myriad fantastic wrecks to dive.

Our discussion turned to the logistics of such a trip. Sam knew from his experience that getting divers to the island with all their equipment could be an expensive nightmare with a dive boat in the area running $1000 per day. If we organized a trip it would be even more logistically daunting as we would concentrate on the deeper and little dived wrecks for which we would have to haul a great number of oxygen and helium supply cylinders, gas booster, and compressor to blend the trimix and nitrox mixtures we would require for our dives.

As Sam pointed out, it would quickly become a very expensive trip for a group of eight to ten divers and could we find enough people locally who were willing to shell out that kind of money for four or five days diving? Good point.

I then mentioned the *Empire Kingfisher*. This wreck could make a great, inexpensive weekend trip and it shouldn't be too hard to find enough bodies to make it worthwhile. Sam and Jason both liked the idea, especially the inexpensive part and knowing that because of it's location on the Province's south shore that it had been little visited by divers.

I already had a lat/long position for the wreck that I had obtained several years earlier from long-time diver Dana Sheppard while we were diving the wreck of the *S.S. Atlantic* together. Dana was one of the few divers I knew who had actually been to the wreck. He had also given me the name of a contact in the area that was a diver himself and would take divers to the wreck.

I told Jason and Sam that I would look into things, try to track down a suitable dive boat and put together a weekend trip to the wreck.

Mike Grebler, Jason Kennedy, and Sam Millett load equipment into the trucks in preparation for diving the Empire Kingfisher.

Author photo.

Being excited by the prospect of diving the *Kingfisher* after procrastinating about it for many years I wasted no time and by that afternoon I had already called Dana's contact in Ingomar, one of the closest ports to where the wreck lay. It had been 10 years since I obtained the position of the *Kingfisher* from Dana so the first thing I asked Bob Welland when I spoke to him was if he still had a boat: he did. And would he be willing to take a group of divers to the *Empire Kingfisher* in August? He would. We had a boat, and a captain who not only knew where the wreck was, but had dived it many times himself.

Now I needed to recruit a group of divers. Not knowing what the response would be, the next day I posted details of the proposed trip on two local diving chat groups as well as sending emails to all those divers I knew who had experience diving wrecks in 160 foot depths and strong currents.

I needn't have worried about finding enough people to fill the ten spots on the trip. One diver sent me his deposit money for the trip only moments after the email was sent out. By the time three days had passed, the trip was full and I had one diver on a waiting list in case a spot opened up. When I booked our accommodations for the weekend I found it would make more sense to put four divers in each of the three cottages I'd booked and so increased the trip number to twelve. This allowed a spot for the waiting diver as well as an extra spot which was also soon filled. By the time the twelve deposits were paid I had four more divers on a waiting list. I guess finding enough divers would be the least of my worries.

I had set the dates for our dives on the weekend of the 4th and 5th of August for two reasons: The first was that the dates coincided with my friend Mike Grebler's vacation when he would be in Nova Scotia for two weeks. I knew Mike would jump at the chance to photograph this wreck. The other, and even more important reason, was that this weekend coincided with the slackest tides in the area that month, meaning we would have less of the area's notorious currents to deal with. Three dives to the wreck were planned, two on the first day and one on the second

The week of the long-awaited trip finally arrived, finding me among my compressor, gas booster, and a ridiculous number of scuba tanks and supply cylinders mixing gas for four of the group's divers. I had elected to have each diver bring all the cylinders they would require to complete their three dives on the wreck, thus negating the need to drag a compressor, gas booster, and gas supply cylinders with us. Even so, it was a difficult task for some of the divers to beg, borrow, or steal enough cylinders from diver friends to make up the up to twelve tanks that each would require. I spent three days, in between other commitments, filling tanks and analyzing gas mixtures for myself, Sam, Jason, and Mike, but by Thursday afternoon all was ready.

On Friday morning, the 3rd of August, Sam, Jason, and Mike arrived at my place and we loaded the many tanks as well as all of our other gear on Mike's truck and mine. We then had a quick lunch prepared by my better half, Janice, and were on the road.

After meeting several of the other divers in Halifax, we convoyed to Ingomar, about a three-hour drive, and found our accommodations at Whispering Waves Cottages, run by Jo-Anne and Paul Goulden. We were happy to see that our cottages were new, clean, and well kept. While several of the divers lit barbecues to cook dinner, three of us undertook the one-minute drive to the government wharf where we could see our dive boat and where we would load our gear the next morning. I had been contacted several weeks earlier by a troubled Bob Welland who because of another commitment that he couldn't reschedule, said he could no longer take us to the wreck, but he had found another boat, larger than his, and a skipper who would gladly take us out.

It was this vessel we had come to check out. We found the *Sea Gypsy III* at her moorings at the Ingomar government wharf, and at 40 feet in length by 17 wide was certainly an able enough boat for our needs. The only thing which gave us pause was the half an aluminum extension ladder lying across her deck: We knew this was to be our dive ladder. We had all used similar ladders in the past and knew we could do it again, but we weren't really happy with the prospect of climbing this less than desirable device while wearing doubles and deco tanks. But that's diving: When using a fishing vessel as a dive boat one can't expect it to be the same as a well set up and diver friendly charter boat. We'd just have to deal with it.

After several months of anticipation, at last the morning of August 4th arrived. The morning dawned grey and foggy, but warm and with no wind. Fog we could deal with.

We were one diver short, as one member had to cancel last minute, and after seeing the ladder we had to get back on the boat, another elected not to dive that morning. The rest of us as well as local commercial diver and friend of Bob Welland's, Donnie Mahaney, loaded our gear onto the *Sea Gypsy* and were off on the one-hour trip to the wreck site.

Commercial diver Donnie Mahaney, complete with double-hose regulator, prepares to dive the Empire Kingfisher. *Old school deep diving at its best!*

Author photo.

Upon arrival near the wreck, Skipper Steve Perry saw what looked like wreckage on the depth sounder. We dropped anchor and because we had the advantage of X-Scooters in the strong current, Harvey Morash and I splashed to do the tie-in. After fighting the current all the way to the bottom what we found at 160 feet was a large, cylindrical boulder that had shown up like wreckage on the boat's depth sounder. We scootered away from the anchor, made a sweep, and then came back to the anchor without finding anything. We began our ascent, meeting Mike, Jason, and Sam at about 90 feet as they were descending. We turned them around and sent them back to the boat; Harv and I had already wasted one dive and already had a considerable decompression obligation, no sense in having more divers do the same.

An hour after we entered the water, Harvey and I completed our decompression, surfaced, and climbed back aboard the *Sea Gypsy*. It's not uncommon to miss a wreck and we weren't the first to miss this one:

Late one night in 2005, I received a call from divers working on a well known undersea television program. They had arrived off Port LaTour (where the wreck is located) to dive and film the wreck for television. They had just spent two fruitless days searching for the Kingfisher. *Would I happen to have an accurate position for the wreck? I thought it strange that they couldn't find such a large shipwreck with their side-scan sonar, ROV's, and other search equipment. But all the high-tech gadgetry in the world wasn't going to find the wreck for them where they were looking: they had been searching five miles away from the actual wreck site.*

This highlights one of the main problems facing shipwreck searchers. Even with modern finding aids if a semi-accurate position isn't available, hunting for a wreck becomes something akin to searching for the proverbial needle in a haystack. And historical wreck positions are notoriously inaccurate.

With the boat relocated and now anchored on what the captain was certain was the wreck, Harvey and I who were on our surface interval, helped the other divers getting suited up ready to dive.

Jason Kennedy hovers above the starboard anchor, still secured in its hawse pipe.

Mike Grebler photo.

Because of the strong current when Harvey and I had made our first foray into the depths, and even though it was now slack water (no current at all), only three teams of divers elected to dive the wreck this time: Sam and Jason were the first to splash, having drawn the short straw this time to do the tie-in. They were quickly followed by Dwayne McLaughlin and Dana Sheppard. Mike and I then helped Donnie Mahaney get geared up. Assisting Donnie was like stepping back in time to the earlier days of diving: He was diving double aluminum 80's of air (for both his bottom gas and decompression), no BCD or wing, and a double-hose Aqualung regulator. Donnie had logged many dives on the *Kingfisher* and on this day he had a fantastic solo dive, unburdened by the extra gear that we "new-school" tech divers carried (certainly not my cup of tea, but to each his own!). Looking like he just stepped out of an episode of Sea Hunt, Donnie jumped from the boat's stern to begin his dive.

Harvey and I then geared up for our second dive of the day. We splashed in, had our scooters handed to us from the boat, and were off.

At about 110 feet I began to see a dim outline below me. This time there was no doubt, we were on the wreck. We arrived on the bow of the wreck at 140 feet. The first thing we noticed was the slab side of the bow with the starboard anchor still secured in its hawse pipe. We circled the bow section and then headed aft. The wreck is much more intact than either the *Kolkhosnik* or the *Kaaparen*, even though it too had been blown open and salvaged. Over the majority of the wreck's length, it stands 20 or so feet proud of the bottom, so that even though the wreck lies at 160 feet, most of it can be traversed at 140, with parts of the bow and stern rising to 130 feet.

As we scootered over the forward cargo areas we were awed by the tens of thousands of three inch ammunition shells, sitting as if still in long-disintegrated crates in both the forward and after cargo areas. Although the ship had been carrying valuable metal cargoes of copper ingots and bismuth, the bulk of her cargo had been these munitions.

We scootered to the port side down near the sand, and into a large, gaping hole in the number two 'tweendeck hold. I knew from my research that this was the area from where the twelve tons of valuable bismuth had been recovered. We shot up through the upper 'tweendeck hatch and back to the level of the ammunition cargo (Harvey later remarked that this maneuver was one of the high points of this dive for him). We continued aft, soon coming to the boilers and triple-expansion steam engine. To gain an idea of how much of this vessel was still intact above the bottom, as we traveled through this area only the very tops of the boilers and engine were visible above the wreckage. This is in contrast to the completely flattened *Kolkhosnik*, whose boilers and engine are of a comparable size and tower up to 20 feet above the bottom wreckage.

We continued through the aft cargo section, once again passing over several thousand 3-inch brass shells, loaded with their charges and projectiles, still patiently waiting to do their part for the Allied war effort.

As we reached the stern-castle area we found that, like the bow, this section was mostly intact and rising up to thirty feet above the surrounding seabed. But where the bow section was lying on its port side, we found that the stern section was leaning drunkenly in the opposite direction, lying partially on its starboard side with the port side of the deck uppermost.

104   Wreck Diving Tales

Above: *Jason Kennedy examines the wreckage of the number 1 cargo hold. Note the three-inch shells scattered beneath the diver.* **Below:** *Jason swims toward the camera in this shot of the forward cargo area. The structure covered in anemones to the far left is one of the cargo handling masts, now collapsed across the wreck. Vehicle parts and ordinance can be seen in the foreground.*

Mike Grebler photo.

We traveled around the stern from the deep starboard side to port, noting the guardrails in some cases still in place and a large stern anchor secured at the very after end of the ship. We continued around to the port side, where the massive rudder and propeller lay exposed with a small vessel's anchor lying in the sand nearby (we were later to learn that this anchor had been lost by Bob Welland and Donnie Mahaney while diving the wreck several years earlier).

Checking our bottom timers, we found that we were nearly 25 minutes into our dive and though we had adequate gas remaining, after our earlier dive we had a long decompression obligation for this one. Harvey signaled me with his light and we both reluctantly agreed to head back, traveling this time along the port side of the wreck as we retraced our path to the anchor line near the bow. After quickly checking the tie in (we would be leaving the anchor and line on the wreck until our dive the following day) we ascended to our first decompression stop at 100 feet, finally surfacing with huge grins on our faces 90 minutes after we began our descent.

Because we had missed the wreck on our first attempt, those divers who had just completed their first dive elected not to dive again due to the lateness of their first descent so we headed back to Ingomar, where we soon had the barbecues stoked up, cold beer in our hands, and tales of our first *Kingfisher* dive on our lips.

The morning of August 5th couldn't have been more perfect if we'd ordered it. The sun was shining out of a cloudless sky and there was hardly a breath of wind. Unfortunately, this morning we had lost two more divers who felt they weren't quite ready to dive to this depth in the strong currents to be found at the wreck site. They headed home that morning without having seen the wreck but promised they'd return next year to make the dives with us after gaining additional experience. Donnie Mahaney, due to other obligations, would also not be joining us this morning.

The eight remaining divers boarded the *Sea Gypsy III* at 10:30 that beautiful morning and headed out to the wreck site. Because slack water that day was expected around 12:30, we elected to leave the dock later to time our dive near slack tide and so minimize the current during the dive.

Arriving at the wreck site, we quickly tied into our buoy left there the previous day. Jason and Mike were the first team in the water, followed by Dwayne and then Dana who were both conducting solo dives today. Harvey and I were next in to be followed by Sam and Kim Langille. This would be a milestone for Kim as it would be her deepest dive to date.

Arriving on the wreck, we found a considerable current that we hadn't felt at the surface. With the X-Scooters this would pose no problem for us, but it would certainly hinder the explorations of the other divers.

Left: *Jason swims over the wreckage of the cargo area. On the right side of the photo can be seen the broken stump of the cargo mast seen in the previous photo.* Mike Grebler photo.
Right: *Jason has his video camera handed down to him from the stern of the* Sea Gypsy *after splashing in for a dive on the* Empire Kingfisher.

Author photo.

Harvey and I began the dive much like our previous one: We scootered down the starboard side of the *Kingfisher*, this time staying near the sand, alert for any pieces of wreckage away from the main body of the wreck. As we approached the boiler/engine room area, I noticed some plating lying in the sand off to the side of the wreck. Stopping my scooter and slowly dropping down near the bottom, I saw small portholes staring up at me from the plate like dead, vacant eyes. I deduced from the thinness of the steel that this piece of wreckage must have once been part of the vessel's midships superstructure, discarded by the salvagers, as the plating was much too thin to have come from the hull.

Left: *Dana Sheppard and Donnie Mahaney, the only two divers aboard who had visited the wreck previous to this trip, discuss their dives on the trip back to Ingomar.* **Right:** *Harvey Morash, Sam Millett, Jason Kennedy, and Mike Grebler discuss their first dives on the* Kingfisher.

Author photos.

We continued on our journey, once again reaching the stern of the wreck, where I made a short and quick penetration into the steering gear compartment, checking out the massive rudder stock and quadrant and the auxiliary steering gear. We once again rounded the stern from starboard to port, and scootered forward once more.

Upon reaching the bow, we found Jason and Mike in attendance. Mike was snapping photos with his trusty Nikonos V camera, using Jason as a model. Mike snapped a photo of me as Harvey and I scootered past the starboard bow and continued out into the sand beyond the wreck. In this area, far from the wreck itself, we found many huge pieces of hull and superstructure plating with many portholes attached. This would have been the salvager's dumping ground for those pieces of wreckage and cargo that were of no use to them. Unlike the dumping areas on the *Kolkhosnik* and *Kaaparen* which were both located off the starboard sides of the wrecks, the *Kingfisher's* dumping ground was located far beyond the bow, and without the scooters, allowing us to cover much more area than the free-swimming divers, we would never have seen this part of the wreck's debris field.

Our bottom timers showing that we had been down for nearly 30 minutes, we made our way back to the anchor and began our ascent. After a further 50 minutes of deco time we broke into the bright sunlight and warmth of the surface, handed our scooters and deco tanks up to those waiting on the boat, and climbed our annoyingly inadequate dive ladder to the deck of the boat.

All hands had thoroughly enjoyed their dives and as this is written I'm already planning a longer trip back to the area for 2008, to dive the *Kingfisher* as well as other area wrecks.

Eight of the nine divers who made it to the wreck of the Empire Kingfisher *on August 4th-5^{th,} 2007.* **L to R, Back Row:** *Dwayne McLaughlin, Dana Sheppard, Kim Langille, Sam Millett, Jason Kennedy, Mike Grebler.* **Front Row:** *Harvey Morash, Al Henneberry.* **Missing:** *Donnie Mahaney.*

Author photo.

History

On July 19th, 1919 spray cascaded into the sky to be turned to twinkling diamonds by the bright sun as a new ship slid down the ways of the G.M. Standifer Construction Corporation in Vancouver, Washington.

The 6038 ton vessel had been built for the United States Shipping Board in Portland, Oregon and was christened *Coaxet*. She was 402 feet in length by 53 feet wide and powered by a single reciprocating steam engine fed by twin oil-fired scotch boilers.

In 1937 ownership of the vessel was turned over to the U.S. Maritime Commission and shortly thereafter was laid up in reserve at New Orleans.

The outbreak of WWII and the advent of the American Lend-Lease program revived several old and nearly forgotten cargo ships including the *Coaxet* and 24 other coal and oil burning steamships. Part of a letter dated September 18th, 1940 from Huntington T. Morse, the Assistant Chairman of the U.S. Marine Commission reads:

> *As a result of advice received that British interests are in the market for cargo vessels, preferably with Scotch boilers and reciprocating engines, 25 vessels of the laid-up fleet have been selected ...*
>
> *As previously reported to the commission, the Navy Department has officially indicated that it would interpose no objections to the sale to foreign purchasers of cargo vessels more than twenty years of age....*
>
> *..... whether or not these vessels should be sold to foreign interests in consideration of the wartime needs of the country at large..... is a matter for determination by the Maritime Commission.*
>
> *..... if some reasonable assurance can be obtained that bids will be received in an amount comparable to the market value of these vessels, there appears to be no valid reason why they should not be offered for sale at this time and thereby permit the Commission to realize prices on vessels which otherwise may very likely continue in lay-up and ultimately have to be disposed of at scrap value.*

And so, the *Coaxet* and her sisters were bought up by the British government and given a new lease on life, to sail for the Allies transporting war cargoes from the New World to the Old under the guidance of the Ministry of War Transport.

The vessels were all renamed as "Empire" ships to distinguish them as British government owned vessels, with the *Coaxet* herself being renamed *Empire Kingfisher* and coming under the management of Crosby, Son, and Company for MOWT. The *Empire Kingfisher* was destined to spend less than a year under the ensign of the British Merchant Marine.

On the 18th of January, 1942 the *Kingfisher* was traveling up the east coast fully loaded with war materials. She was on her way to Halifax to join a convoy to Britain and had just made contact with the southern coast of Nova Scotia when she struck a partially submerged object. It was later suggested that this was a vessel which had been sunk earlier but was still floating just beneath the surface although there has been no definitive proof of this. In any event, whatever the object was it had holed the *Kingfisher* causing flooding and ultimately the failure and shutdown of her propulsion machinery.

With the ship dead in the water, the captain ordered the port anchor dropped and the vessel came to rest just outside the entrance to Port LaTour. The armed yacht *HMCS Lynx* removed the *Kingfisher's* passengers and crew and called for a salvage tug from Halifax to tow the now derelict freighter to a safe port where she could be repaired.

Late on the night of the 18th the German submarine *U-109*, commanded by Kapitanleutnant Heinrich Bleichrodt came upon the stopped and abandoned *Empire Kingfisher*. Bleichrodt couldn't believe his good fortune as he lined up the cargo ship in his periscope in preparation for launching a torpedo at the sitting duck.

The Empire Kingfisher *while still under American control as the* Coaxet.

Photo courtesy of Hubert Hall, Shipsearch Marine.

At just a few minutes before midnight local time from a range of 800 meters, Bleichrodt fired his first torpedo. It didn't explode. He fired a second one from the same spot which also failed. He then moved in to 500 meters before firing his third fish at the ship. When this one also failed to explode and sink the unmoving vessel, he moved off to 1500 meters to fire the fourth torpedo. Again no hit. He once more moved in to 500 meters, determined to sink his quarry. Using his net guard as a cross hair to line up the shot, he ordered the fifth fish to be fired at the *Kingfisher*. Another failure! In disgust, Bleichrodt turned his bow away from the still floating vessel and steamed away.

This incident highlights the problems that the German U-boat fleet had with the exploders as well as the depth sensors on their G7e torpedoes which were the weapons the *U-109* was equipped with at the time. This scenario was reported time after time by many u-boat skippers to u-boat command who early on blamed the captains and refused to believe there was a problem with the weapons. After continual reports of failures however, the exploders and depth sensors were finally re-engineered and the balky weapon became the deadly fish which sank hundreds of thousands of tons of Allied shipping.

After the departure of *U-109,* the *Empire Kingfisher* slowly settled and sank upright to the bottom in 160 feet of water on the morning of January 19th before a salvage tug could reach her. Bleichrodt could perhaps take some solace in the fact that even though he was unsuccessful in sending the vessel to the bottom, she did eventually sink and he was given credit for the sinking after the war.

In 1952 the salvage vessel *Help* was sent from England by Risdon Beazley to recover the *Empire Kingfisher's* part cargo of copper and bismuth.

Captain Tom Young of the *Help* anchored his vessel on a six point moor and began preparations to blow open and salvage the wreck. Because the depth of the wreck was not too great, the salvagers were able to use hard hat divers to lay the charges used to open up the wreck, something that wasn't possible with the deeper wrecks they worked.

After divers Sam Dooley and Bob Mouncer laid the charges, the salvage vessel was moved off the wreck site and the charges detonated to open the ship allowing access to the cargo. With this done the *Help* was brought back on site and the observation bell was lowered to the wreck site to direct the recovery operations. But before it could begin recovering cargo, the huge eight-legged grab had to remove tons of deck plating and many more tons of worthless cargo before it could begin picking up the valuable copper and bismuth which was stowed in the lower holds of the ship.

Eventually, 499 tons of copper ingots and 12 tons of the valuable bismuth were recovered from the *Empire Kingfisher*. When the grab could no longer pick up the small ingots of bismuth, the hard hat divers were sent back down to collect all that they could find and place it in canvas sacks.

When the recovery was complete, the metals were landed in Shelburne, trans-shipped to the tug *Turmoil* which had been chartered by Risdon Beazley to carry the cargo back to the UK, finally ending for the metals the journey that had begun in 1942.

With the *Kingfisher* job finished, Risdon Beazley had concluded their first salvage of a war wreck in Canadian waters. The salvage crew then steamed up the coast to Halifax to begin salvage operations on the *Kolkhosnik*.

CHAPTER 11

Halifax Harbour Diving: Bottles & China

In September of 1746, the Duc D'Anville arrived in the harbour of Chebucto with the remains of his battered and sickly fleet. The plan had been to join forces with another French fleet coming from the West Indies and to take back Louisbourg and drive the British from the North American continent once and for all. But things didn't quite work out as planned.

D'Anville's fleet was decimated by storms and scurvy by the time the remnants reached Nova Scotia three months after leaving France. D'anville himself died shortly after the fleet's arrival, leaving a man named Jonquiere in command. The expedition had lost over 2400 men due to storm and disease and it was obvious that re-taking Louisbourg was now impossible. Jonquiere burned several of the fleet's unseaworthy ships and left the Harbour on October 13th to return to France. In his wake, the waters and shoreline of the beautiful inlet were littered with the bodies of hundreds of his countrymen, killed by scurvy and typhus.

Three years later, the English arrived. The president of the Board of Trade and Plantations in London, Lord Halifax, drew up a proposal for a settlement to be started in Nova Scotia. The western shore of the harbour at Chebucto was decided as an easily defendable position, and so Colonel Edward Cornwallis and almost 2700 settlers set sail from England in May of 1749 in the sloop of war Sphinx.

The beginnings of this venture were rocky at best, as most of the "settlers" were soldiers who knew nothing of living off the land, farming, and starting a new community so far from home. But from these modest beginnings, the town prospered and eventually grew into the city we know today.

Through the years, as the city grew, numerous businesses and shipping interests grew along with it. Among these ventures were several soda water manufacturers, ginger beer and beer breweries. In addition, Halifax was ever a garrison town, seeing not only the British Royal Navy's ships, but also the ships of other navies of the world. Not to mention being a convoy mustering point and troop departure point through two world wars, as well as a great merchant shipping port.

The civilian population too, greatly expanded. The town spread out, becoming a city which eventually spread out to areas other than the easily defensible peninsula on which it was originally founded. In the late 1800's, the Northwest Arm of the Harbour became a place of leisure activities, host to regattas and sailing races. Boat traffic of all types became quite concentrated along this narrow waterway. Point Pleasant became a favorite Sunday destination for picnics and family outings.

One result of this growth in the city, from its earliest days right up to the present, is that Halifax Harbour has become the repository for enormous amounts of detritus thrown into it by its residents and also by the crews of visiting ships of all kinds. Much of this waste was of course biodegradable and would have broken down in a short time. But other items, such as bottles, china, and in some instances even shipboard items such as binoculars and officer's swords, have remained. Many of these artifacts have of course been smashed and destroyed by the harsh environment in which they reside, but many others have been preserved by the soft sediment of the harbour bottom and are eagerly sought, some in pristine condition, by divers.

Areas such as the Narrows, the Halifax Dockyard, the Northwest Arm, and Pier 21 have turned up treasure troves of artifacts over the years and these spots are enthusiastically searched by divers each winter.

An Introduction to Bottle Diving

Although primarily a wreck diver, I too have become enamored of this winter pastime. Because of our climate, adverse weather conditions make wreck diving in Nova Scotia a hit or miss proposition in wintertime. Whereas diving in the Harbour can be undertaken throughout these months of unfavorable weather because of the sheltered condition of the Harbour and the fact that much diving can be conducted from shore. The only thing left to contend with is the water temperature, which often drops to a bone-chilling 29 degrees Fahrenheit around February and March. This is offset by diving using a drysuit with heavy undergarments and using argon for a drysuit inflation gas, its density providing greater insulation than regular air. For those of us who spend much of the year diving shipwrecks in the open ocean, harbour diving allows us to keep our skills sharp during those months when we otherwise would not be able to dive.

By the time I was introduced to bottle diving, there were already several groups of divers in the area who had been diving Halifax Harbour regularly for more than twenty-five years for old ginger beer and other bottles. My introduction came on December 2nd, 1998, when I received a call from my friend Sam Millett who at that time worked for Dive Masters Scuba Shop in Dartmouth.

"We're going for a night dive after I get off work, are you coming?" Sam asked.

"Where?" was my cryptic reply.

"Fleming Park on the Northwest Arm."

At this point in the conversation I began to wonder about my friend's sanity and asked why anyone would voluntarily subject themselves to the filthy waters of Halifax Harbour. This time it was Sam's turn to answer with one word: "Bottles."

Or more specifically, antique bottles which have been tossed into the harbour on a regular basis for more than 250 years.

On that first dive I found a small, clear glass bottle with a ground-glass stopper in it and after Sam confirmed that it was indeed a good find, I was hooked. The bottle also contained some vile smelling oily liquid which I sent to have analyzed at Dalhousie University. When I told Dr. Robert Ackman that I suspected the foul, viscous substance was whale oil, he seemed as enthused as I. When the results came back though, it was nothing as glamorous as whale oil from a Quaker whaling ship used to light a lamp in Halifax's early days, but only hundred year old rancid olive oil.

The next night found us back at the Arm, on the opposite side, at the foot of South St. This time I had brought along my good friend and dive buddy Paul Crocker. Although my luck wasn't as good as it had been the previous night, there was something about this search through Halifax's past that appealed to my historically inclined nature and on December 23, we were at it again. All through that winter we kept diving the Northwest Arm, periodically coming up with decent finds, but mostly just refining our search techniques. Usually it was Paul Crocker and I who conducted our dives together, veering off from other groups and heading off on our own to search.

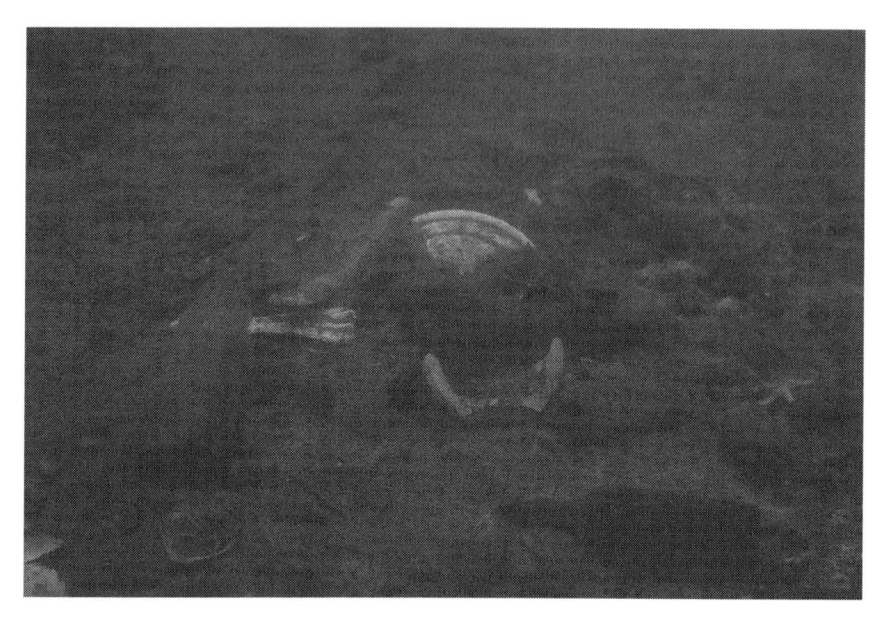

The harbour bottom at Pier 21. A few bottles, a broken coffee cup, and other debris can be seen. This photo gives an idea of what much of the bottom of Halifax Harbour looks like.

Author Photo.

Before getting to the following winter when our bottle diving began to bear fruit, I would like to take a moment to speak of the environment of Halifax Harbour itself. For more than 250 years the effluence and sweepings of a growing city have been discharged into Halifax Harbour. To anyone who has never seen the bottom of the harbour it stands to reason that this dumping continues down to the harbour bottom. This is only partially true: while heavier objects and some sediment will settle to the bottom, much is swept away on the twice daily tides running into and out of the harbour. Also, the trash and oil often seen floating in the water of busy seaports (and Halifax is no exception) doesn't make it to the harbour bottom and drifts away on the tide or washes ashore. The point I'm making is that although Halifax Harbour is by no means clean, it's not nearly as filthy as the average citizen would think. There are certain portions with a strong tidal flow that are almost pristine.

Pier 21

One day in the fall of 1999, I found myself at Torpedo Rays Scuba Ventures in Dartmouth, speaking with my friend Tye Zinck, who worked there as an instructor and equipment tech. The topic got around to harbour diving and I was invited to a new spot: Pier 21 on the Halifax waterfront.

From 1928 to 1971 Pier 21 was the entry point for thousands upon thousands of immigrants coming from Europe to make a new life in Canada. It was also the departure point for troopships leaving Canada during World War II and returning with the fighting men at the end of hostilities. This made it a perfect place to search for shipping line china and other artifacts and indeed several divers had already had tremendous luck at the site.

On November 21st, 1999, Paul Crocker and I pulled in to the Pier in my truck which was loaded with our dive gear. Several other divers, among them Paul Grantham, Tye Zinck, and Jared Rainault, were already putting their gear together near their vehicles on the Pier deck.

Paul and I wasted no time getting our gear ready as well and entering the water with the other divers. Once on the bottom at the foot of the Pier, we headed away from the others. The depth was 45 feet and sloping downward away from the Pier. There were also huge craters in the bottom made by the thrust of cruise ship's propellers. The cruise ships which still visit Pier 21 have a double edge sword effect on the artifacts lying on the bottom: Because of the silt in the area, hundreds of artifacts are buried in the deep mud and would never be seen by divers if not stirred up by the turbulence created by these ships, to settle gently back on top of the silt where divers can find and reach them. The other edge of the blade is that the force produced by the ships is so great and the water so agitated, that hundreds, if not thousands more artifacts have been smashed and destroyed.

This was evidenced as we swam along by hundreds of broken plates, cups, bottles, and shards too small to identify.

Still, by the end of the dive our goody bags were getting heavy and we each had made a decent haul with several plates bearing shipping line logos (a couple famous ones like Cunard's and White Star Lines), as well as a few bottles and in my case, an "Old Spice" shaving mug from the 1940's. All in all, it had been a successful dive and a good start to our winter diving activities.

After a few additional dives at the Pier, mid-January found us back at South Street on the Northwest Arm, where we continued our diving with varying

Above: *Artifacts recovered from Pier 21.* **L to R:** *A French mustard jar, butter dish, Canadian Pacific Lines egg cup, Cunard Steamship Company egg cup, hand painted porcelain noodle bowl, Dempster Lines coffee cup.* **Below:** *Exhibit housed in the Pier 21 Museum Research Center containing artifacts recovered during the 2006 dives at Pier 21.*

Author photos.

luck until the weather broke and we could once again head out to our favorite wrecks. By this time almost every dive was yielding bottles which we considered keepers: soda water bottles from Halifax companies such as James Roue and Felix J. Quinn, as well as milk bottles from various local companies, and medicine bottles embossed with the names and quack remedies of local as well as American druggists. Our collections were growing and bottle diving was gaining an ever stronger hold over us.

After the attacks of September 11th, 2001, security in the Port of Halifax was systematically tightened. By 2002, sport divers were no longer permitted to conduct dives from Pier 21 or in the waters adjacent to it.

In the summer of 2005, before his departure to Ottawa, Mike Grebler donated some of his Pier 21 finds to the Pier 21 Museum. After being told about this, I decided to do the same as I had stacks of doubles of many of my Pier 21 artifacts and thought the museum might be able to use them. While visiting the museum to make my donation, I was met by Carrie-Ann Smith, the Pier Museum's research librarian. After discussing the artifacts and what might still be lying in the waters near the Pier, we talked about the possibility of conducting dives for the purpose of collecting artifacts for the Museum.

The Port of Halifax gave its blessing and on January 15th, 2006, Jason Kennedy, Sam Millett, and I undertook the first of three dives that winter to collect artifacts for the Museum. In addition to the many interesting artifacts we recovered, Jason shot underwater video of one of the dives and I shot still-photo documentation. All in all, the dives were a great success and plans were made to return in the winter or spring of 2007.

Message in a Bottle

As with the previous autumn, the fall of 2000 found us back in the Northwest Arm and much earlier than before. On September 25th, Paul and I were at South Street and I found my first stone ginger beer bottle, a "James Roue 1903". On the same dive I picked up the first of many clay smoking pipes.

I was to continue picking up clay pipes on a regular basis, a happening which caused Paul no end of frustration. I would finish a dive and show off my finds, often including a pipe or two, while Paul remarked on how he could never find any and how the hell did I manage to see them? Eventually he would begin spotting them, but remained stymied throughout that season.

It was during the fall of 2000 also that we happened upon a new search method. We discovered that about 6–18 inches of silt at this point in the Arm was lying atop a harder gravel bottom. Any bottles which sank in the Arm would therefore sink through this soft layer and rest on the hard gravel below. We began swimming along the bottom with our hands and arms extended into the silt like ploughshares, feeling our way along the buried gravel and picking up bottles as we felt them. It worked! We began picking up as many as four ginger beer crocks each per day.

By this time we had been joined in our activities by neophyte diver Jason Kennedy teamed up with long-time diver and friend Mike Grebler. The four of us became a tight-knit group and conducted almost all of our diving activities together for the next five years until Mike relocated to Ottawa.

The result of four of us dredging the Northwest Arm in this manner is that hundred year old trash (today's artifacts) happens to be a non-renewable resource and by the end of that winter ginger beer crocks were becoming a scarce item in the Arm. Try as we might, we couldn't think of a way to seed the harbour bottom so it would grow more: we would have to move on to greener pastures.

Besides the ginger beer crocks and other soda bottles that winter, I did have one other peculiar find: a bottle with a note in it. On December 8th, Paul and I were diving at Fleming Park, back once again on the Arm's western side. I had picked up a small medicine bottle during the dive which, because it wasn't embossed, I would normally have tossed back. It did however, still contain its cork, so I tossed it in my bag and continued on. At the end of the dive while we were showing each other our finds, I pulled out this small bottle. Looking at it, I noticed paper inside and remarked to Paul that it seemed to contain a note. I left it alone for the time being, but when I arrived home I extracted the note and carefully tweezed it open. Upon the small slip of paper, written in pencil, was:

Miss Margie MacMillan
Halifax,
N.S.
38 South Clifton Street
July 24th/15

———

Miss Helen Shaffelburg
65 Charles St.
Halifax, N.S.
July 24th, 1915

I carefully preserved the note and mounted it, along with its bottle, on a small varnished plaque.

I later decided to see if I could make contact with a relative of either of these girls. I decided that MacMillan was a long shot as it's such a common name and so would concentrate on Shaffelburg. As it turned out, there were three listings in the Halifax phone directory, so on a whim I picked the middle number and called it. An elderly gentleman answered and after I explained why I was calling, he told me that he had indeed known Helen Shaffelburg, that they had lived in separate flats in the same building at 65 Charles Street. Although both possessed the same uncommon last name and lived in the same building, they were not related, and he lost touch after her family moved away in the 1920's. The other two listings in the phone book were this gentleman's son and grandson

What I did learn from him was an approximate age of the girls, leading me to surmise that they were around fifteen at the time they wrote their note. My deduction is that they had intended their bottle to drift away on the tides, perhaps to some foreign, exotic land where the discoverer of the bottle would find their note and write to them. However, they didn't finish their homework regarding the vessel in which their message was to be transported. When I tested the empty bottle for buoyancy, it immediately sank, the airspace in it being too small to float the weight of the glass it contained. When they sealed the bottle and tossed it into the waters of the Northwest Arm on that summer day in 1915, it immediately sank to the bottom only to be found by a diver 85 years later. Rather than traveling to an exotic land, their note traveled forward to a time they could only imagine while remaining in precisely the same place.

The small bottle and message it contained recovered from the Northwest Arm adjacent to Sir Sanford Fleming Park on December 8th, 2000 by the author.

Author photo.

Angus L. MacDonald Bridge

During the winter and spring of 2001, we began diving the area of the Harbour adjacent to the Naval Dockyard and near the MacDonald Bridge. This proved to be a premium spot for this had been the busiest part of the harbour since the founding of the city and also had the advantage of being swept clean of silt by the tides which rush through this narrow part of the waterway.

In this area we began finding items which were much older than any we had recovered previously: free blown black glass bottles dating back to the 1750's, stoneware ink wells, as well as older ginger beer crocks, and many bottles and items from other countries.

Although this area was rich with artifacts, it was a difficult spot to dive from shore. We couldn't enter from the Halifax side due to access to the dockyard being prohibited, and it was a long swim from the Dartmouth side through water depths up to 90 feet to the area which was producing the best artifacts. The currents in the area also meant that this was a difficult spot to dive at any time other than slack water. We returned to the Northwest Arm for our bottle dives, periodically trying other promising spots along the Halifax waterfront, but finding none which produced the way the area near the dockyard did.

We decided the area adjacent to the MacDonald Bridge was the place to be, but the dives we wanted to do would be too long to be conducted with single tanks from shore. Although we always used double tanks while wreck diving from a boat, the often long walks while wearing full gear, as well as the treacherousness of some entry points made doubles prohibitive for bottling. With a long swim to the most productive area and depths in the range of 70–90 feet, we bit the bullet and decided doubles were the only way to go (consequently I no longer dive single tanks on any of my dives).

Our first dives from the Dartmouth shore with doubles in the spring of 2002 were unqualified successes. We were extending our bottom times to over an hour and coming back with more artifacts from an earlier period in the city's history than at any time previously. A late season dive, on June 27th of that year, is a case in point.

Paul, Jason, Duane Williams, and I (Mike was indisposed on this day) entered the water and swam out past the first bridge abutment near the Dartmouth shore where the depth dropped off quickly from 35 feet to 70. As we scanned the bottom I quickly spied a stoneware marmalade crock which I placed in my goody bag. Continuing along, I located a ginger beer crock, a stoneware inkwell, then another marmalade crock. And so it continued for the remainder of the dive, with a Codd stopper bottle rounding off my haul of fifteen keepers while on our swim back to Dartmouth's rocky beach.

Just before the bottom began to slope upwards, alerting us to our position near the Dartmouth shore with only about a seven minute swim remaining, I spied a fire-hydrant shaped, dark object partially buried in the silt. I tugged it free and saw it was made of glazed stoneware, but still had no idea what it was other than being old. Having no room left in my goody bag, I carried the heavy item in my right hand, adding air to my BCD to compensate for the extra weight. After a grueling ten minute swim which was more a crawl due to the weight of the object and my full goody bag, I arrived at the shoreline along with my fellow explorers.

After de-kitting, we returned to the shore to clean and catalogue our finds. It was then that we discovered that what I had picked up was a Victorian-era charcoal water filter, made of glazed stoneware and ornately decorated. After cleaning, it would be put on display in my kitchen and remains one of my favorite finds.

A final note about this main area of Halifax Harbour: The divers who frequent the harbour bottom in the area of the Narrows occasionally come upon pieces of ship wreck. Now, given the history of Halifax Harbour such a "discovery" would likely be viewed as mundane, even expected. But, there's something different about these pieces of wreck. These are clearly old; they're riveted iron instead of welded steel, twisted and heavily corroded. You can see that a cataclysmic force was involved in bending steel plate against the rivet heads and causing elongated tears in the metal, almost like shards of glass. It takes a few moments but then you begin to comprehend where you are and what happened here so many years ago. You closely examine the wreckage and run your gloved hand along a jagged edge. In the quiet darkness, a cold chill passes over you. There's no possible way to prove to anyone that this was a piece of the *Mont Blanc* (See Chapter II). But, at the same time, you'd defy anyone to put their hand on the same piece of wreckage and tell you—its not.

Our adventures diving in Halifax Harbour have continued each winter since, when the weather turns savage and our wreck diving season grinds to a halt. Although not the type of diving to be displayed in the pages of dive travel magazines or to draw huge numbers of tourists to Nova Scotia, for a few groups of hard-core divers who brave the sub-zero temperatures and fierce winds of winter, it has become a pleasant pastime to conduct these excursions, delving in their own way into the rich history of the busy seaport of Halifax.

The charcoal water filter recovered by the author near the MacDonald Bridge. The inscription on the front reads: "Lipscombe & Co. PATENT 233 Strand, Near Temple Bar, LONDON".

Author Photo.

Above: *Free blown mallet bottles recovered from the Narrows.* **Below:** *Part of the author's collection of ginger beer bottles recovered from various dive sites around Halifax.*

Author photo.

Left: *A free blown black glass jar recovered from the Dockyard area of Halifax Harbour. This bottle is still corked and contains its original contents which can be dimly seen through the glass when held up to a bright light and appears to be olives.* **Right:** *A pair of French binoculars manufactured by LeMaire of Paris recovered by the author in the Narrows near the Angus L. Macdonald Bridge. Originally the entire brass artifact was covered with the black enamel seen on the right side but when recovered this was all that remained of the coating.*

Author photos.

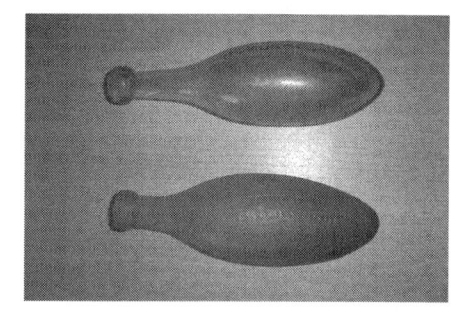

Two torpedo bottles from the author's collection. The upper is a J.B. Heyl's Soda Water *bottle from Hamilton, Bermuda. The lower is an* H.W. Glendinning, *one of the earliest and rarest of the Halifax soda water bottles.*

CHAPTER 12

▼

DEEPER INTO THE ABYSS

The previous chapters give some indication of the variety of wreck diving to be undertaken off the Nova Scotia coast, but the story doesn't end with the myriad known wrecks along our shores. Intrepid adventurers and wreck divers are forever searching for new wrecks, new sites to dive, and there are many that immediately come to the mind of technical wreck divers, both here at home and those from far away.

Gordon Fader, now retired from the Bedford Institute of Oceanography, has been at the forefront of most of the recent discoveries. While conducting surveys of the approaches to Halifax Harbour, Gordon and the Geological Survey of Canada charted over 50 anomalies believed to be shipwrecks, many of them previously unknown. Wrecks such as the *Kaaparen, British Freedom, L-26,* and *Halfish* were all given accurate positions due to Gordon's efforts, thus making them easily located by divers.

Most of these new finds have been in depths previously viewed as unsafe or unreachable by scuba divers, but with mixed gas and rebreather training now being readily available in the area, not to mention more reliable equipment and techniques that are now available to divers, these previously unreachable wrecks are now diving possibilities and indeed divers have already discovered and dived a few of the deeper ones. Others continue the search for elusive wreck sites known to exist but which remain hidden due to inaccurate positions taken at the time of the vessel's demise.

Here are just a few of those deep water wrecks that have already come to light or are now being sought.

Athelviking

The steam tanker *Athelviking*, official number 162387, began life in 1926 as the *Java* for A/S J. Ludwig Mowinckels Rederi of Bergen, Norway. The tanker was 8875 gross registered tons, 471 feet in length, 62 feet 6 inches in breadth, and had a draft of 27 feet 3 inches. She was built as hull number 86 by the Furness Ship Building Company in Haverton Hill, Stockton on Tees, England. J.G. Kincaid & Company had built her twin oil fired steam engines, each producing 709 horsepower and pushing the tanker through the water at a top speed of 10.5 knots.

In 1933 the *Java* was purchased by the United Molasses Company and had her name changed to *Athelviking*. In 1940 the ship was transferred to Athel Line Limited, a company who were well known at the time for their specialty of transporting bulk molasses.

As mentioned in Chapter V, the *Athelviking* was one of the three ships of Convoy BX 141 to be sunk by the U-1232 on the morning of January 14th, 1945. At the time she was under the command of Captain Egerton Martin and had just completed a run up the coast from Fort Lauderdale, Florida with 11,360 tons of molasses in her tanks and a cargo of landing craft on her added spar-deck when she joined BX 141 in Boston for the final run to Halifax.

Captain Martin, his first mate Geoffrey Ince, Bosun Bertram MacHale, and officer's steward Eric Carlsson were all killed when the ship was torpedoed. The 39 remaining crewmen as well as the ship's complement of 8 naval gunners were all picked up by *HMC Motor Launch 102*.

The *Athelviking* lay undisturbed in the dark, cold waters for 58 years. In the summer of 2003, Greg Mossfeldt was leading his second technical diving expedition to the Halifax area. One of the team's goals on this trip was to dive the *Athelviking*. On day seven of the trip, July 24th, Skipper Dave Gray dropped the anchor onto the wreck of the tanker, sitting upright and intact in just over 300 feet of water and reaching up to 270 feet at her bridge deck.

That first day only three of the team, Bryan Eaton, Charlene Barker, and Mark Gangl were able to dive the wreck. But before the expedition was completed, they were able to get back to the tanker twice more and the entire team of six divers got to dive the wreck. These were the first divers to visit the *Athelviking* and are still among only a handful of divers to have seen this amazing wreck.

The Athelviking, *shown here during wartime with guns mounted on the stern and spar-decks added over her tank decks to carry vehicles or other cargo.*

Author's collection.

H.M.C.S. Clayoquot

HMCS Clayoquot, pennant number *J-174,* a Bangor class minesweeper, was launched from the Prince Rupert Dry Dock & Shipyards of Prince Rupert, British Columbia on October 3rd, 1940. She was a product of the 1939–40 build program and as such was built as a coal burning steamer. She was 180 feet long by 28 feet wide and 672 tons. She could travel at a top speed of 16 knots and was armed with one 4-inch, one 3-inch, and two 20mm guns.

On December 24th, 1944 the *Clayoquot* was part of the escort group for the Boston bound convoy XB-139. While the convoy was forming up outside Halifax Harbour for the run to Boston, *U-806* fired torpedoes damaging a ship in the convoy. *Clayoquot,* along with *HMCS Kirkland Lake* and *HMCS Transcona* began a search for the u-boat, and a short time later, at 1:37 pm, *Clayoquot* herself was struck near the stern by an acoustic torpedo from *U-806* and sank within ten minutes. Seventy-six of her eighty-four man crew were picked up shortly after by *HMCS Fennel,* but four of her officers and four seamen would never come home to spend Christmas with their families.

HMCS Clayoquot.

Paul Grantham collection.

Enter once again Gordon Fader. In 1994 his team imaged a wreck outside Halifax believed to be that of *HMCS Clayoquot*. In 1996 the identity was confirmed when a manned submersible from *HMCS Cormorant* visited the wreck. On board the submersible for the dive was Ernie White, a retired petty officer and survivor of the *Clayoquot's* sinking, returning to the ship that had been sunk out from under him fifty-two years before.

In 2001 Mike Fletcher conducted a dive on the wreck of the *Clayoquot* for the television series The Sea Hunters episode *Leopoldville/Clayoquot*. Although Mike's dive was conducted utilizing surface-supply diving gear instead of scuba it still makes him the only swimming diver to visit the wreck to date. The dive was carried out from Skipper Dave Gray's dive boat *Ryan & Erin*.

Although there has been talk in the technical community of diving the wreck, as of this writing no dives have come to fruition. The wreck lies in 330 feet of water.

H.M.C.S Esquimalt

HMCS Esquimalt (pennant number J-272), like the *Clayoquot,* was a Bangor class minesweeper. But being from a later build program (she was launched in August of 1941) she was powered by diesel rather than steam propulsion. The *Esquimalt* was somewhat smaller than the *Clayoquot* as well, being only 162 feet in length by 28 foot breadth and 592 tons. She was built by Marine Industries Ltd. of Sorel, Quebec.

The *Esquimalt* had the dubious distinction of being the last Canadian warship to be sunk by enemy action in World War II.

Three weeks before Germany's surrender, Admiral Doenitz' u-boats were still very active off the east coast of North America. The *Esquimalt* was sent out on the evening of April 15th, 1945 to make a routine anti-submarine patrol in the approaches to Halifax Harbour. Lurking in the vicinity was Oberleutnant Hans-Erwin Reith in the *U-190*. The pinging of the *Esquimalt's* Asdic could be heard but the submarine was not detected. Reith brought his submarine to periscope depth and observed the *Esquimalt* a quarter of a mile away. At 6:30 am on April 16th, the *U-190* fired an acoustic homing torpedo at the *Esquimalt.* The tin fish took the minesweeper in the stern, tearing a huge hole in her starboard quarter. The *Esquimalt* sank in less than four minutes, carrying some of her crew with her to the bottom. Most of the crew had gotten off the ship and were huddled together on four carley floats.

Esquimalt was to rendezvous with *HMCS Sarnia* that morning at 8:00 west of Chebucto Head. By that time most of the men, being soaked through and lightly clad, were suffering from exposure and many had died since the attack. It was after 11:00 before it was realized that something had indeed happened to *Esquimalt* and a search was begun. At 12:30 *Sarnia* located the survivors and picked them up. Few of them could walk by this time and had to be assisted below decks. Only 27 of the 71 men aboard at the time survived their ordeal.

Only a rough position for the *Esquimalt* sinking had been recorded at the time and it wasn't until a naval survey in the late 1950's was conducted to record magnetic anomalies where submarines might hide themselves that a position was pinpointed. But with the electronic navigation systems in use at the time (Decca and Loran A) the position was still far from accurate. The *Esquimalt* has yet to be found but it is somewhere near the eastern end of the Alpha shipping lane heading into Halifax. The wreck is believed to lie in 300 plus feet of water.

U-190 & U-889

U-889

By the time the *U-889* left Horten in April of 1945 for her action station off the East Coast of North America, the war was all but over. The German war machine was grinding to a halt and Hitler's thousand year Reich was looking like it would last only a small fraction of that. The Allies had done their job well. Convoys from North America had for the past 6 years flowed across the Western Ocean carrying in their bottoms the materials that would win the war and allied troops were forcing the Germans back to their homeland. But still the u-boats went to sea to do their part, even though at this late stage the life expectancy of their young crews could be measured in mere months.

And so it was that Kapitanleutnant Friedrich Braeucker took *U-889* to sea on her first (and subsequently last) war patrol, sailing on April 6th, 1945. His orders assigned him to Quadrant BB, off the northeast coast of North America, to sink coastal shipping and harry the convoys sailing up the coast from New York and Boston and forming up in Halifax for the trek across the Atlantic to Britain.

U-889 was a type IX-C/40 u-boat. She was built by Deutsche Schiff und Machinenbau AG in Bremen. The IX-C/40 was a modification of the type IX-C with a slightly increased range, higher surface speed, and a slightly greater displacement when submerged. She was 250 feet in length and 22 feet 4 inches in breadth. She was powered on the surface by twin diesels producing 4400HP which could propel her at 19 knots and underwater by two 500HP electric motors driving her at a maximum speed of 7.5 knots. She could dive to depths of nearly 800 feet to escape enemy destroyers and bite back with 4 bow and 2 stern torpedo tubes for which she could carry 22 torpedoes, with 10 being stowed in deck canisters outside the pressure hull. She was also equipped with a deck gun and anti-aircraft guns, as well as a Schnorkel array with an anti-radar reflective covering, allowing her to run her diesels while submerged to charge her batteries. To keep the vessel operating and fighting, she carried a crew of between 48 and 56 men.

The submarine had been launched the previous year, on April 5th, 1944 and was subsequently commissioned into the Kreigsmarine on August 4th. Upon her commissioning, she was assigned to the 4th U-Boat Training Flotilla in Stettin and remained there for her sea trials and crew training until March 15th, 1945 when she joined the 33rd flotilla at Flensburg.

Even with all of Germany's latest technology crammed within her pressure hull, *U-889* never sank a ship. She was still some distance east of Newfoundland

and heading toward the approaches to Halifax when Germany surrendered to the Allies. On May 10th Braeucker received the order from Admiral Doenitz that all U-boats still at sea were to make their way to an Allied port and surrender. Braeucker surfaced the boat and was over flown by an RCAF Liberator of 10 Squadron. The Liberator was preparing to depth charge the submarine when they saw a flag being flown in surrender. The Liberator circled the U-boat until an escort group comprising *HMCS Rockcliffe, Dunvegan, Saskatoon,* and *Oshawa* arrived to escort the surrendered sub to port. The following day she was handed over to *HMCS Inch Arran* and *Buckingham* who led her into harbour at Shelburne, Nova Scotia. The articles of surrender were signed on May 13th, formally ending *U-889*'s employment with the Kreigsmarine.

U-889 was duly commissioned into the Royal Canadian Navy as *HMCS U-889* on May 15th, 1945. An RCN crew was detailed to the u-boat for the purpose of conducting trials. While conducting diving trials the crew discovered that the boat was very tender and could not be taken down at a steep angle for fear that she would enter into an uncontrollable dive. By all accounts, the crew of hardened RCN submariners who crewed the boat were fearful of diving it and were only too happy to hand the boat over to the American Navy for further trials in January of 1946.

The Americans were particularly interested in the waffle-like rubber covering on the Schnorkel assembly which absorbed rather than reflected radar impulses. This seems to have been the U.S. military's introduction to and the beginning of their own research into Stealth technology.

The story of *U-889* then becomes muddled. All contemporary accounts say that the submarine was sunk by the US Navy during torpedo trials off New England in late 1947. There are a couple of problems with these accounts. First and foremost is the fact that by 1947 the Americans knew full well how their torpedoes operated and also of the flaws to their magnetic exploders during the Pacific war which had since been overcome. So torpedo trials would hardly seem necessary. The other problems with this story are eyewitness accounts of the sinking of the U-boat near Halifax in 1949.

A de-commissioned submarine was reportedly tied up in Halifax awaiting the convenience of the ship-breakers. Research conducted by myself and a fellow technical diver led us to believe that this was in fact *U-889*, returned from the U.S. after they learned all they could of her secrets and had no further use of the u-boat.

The gunnery range on Osborne Head in Cow Bay was looking for a floating target when it was suggested that this derelict sub could be towed to the Target

Corridor and used for target practice. A subsequent eyewitness account from the son of one of the Osborne Head gunners was obtained. In 1949, he was a lad of five, but remembers watching as the submarine was shelled and sunk by the Osborne Head battery. A retired Canadian Navy diver remembers diving on the sub in the mid-70's. At the time when the price of silver was at its highest, he conceived of salvaging the u-boat's banks of batteries for the silver they contained. When the price of silver dropped, he considered the venture to no longer be profitable and so abandoned his idea of salvage. Thus was the *U-889* forgotten and through the years presumed sunk by the USN.

The submarine has still not revealed its resting place to Halifax's small cadre of technical divers. Plans are in the works however to conduct a search for *U-889*, believed to be resting in 180–200 feet of water near Halifax.

U-190

U-190, like U-889, was a type IXC/40 u-boat, built by AG Weser of Bremen, launched on June 3rd, 1942 and commissioned into the German Kriegsmarine on September 24th, 1942. Unlike the *889*, this boat was an old veteran of the Atlantic war, although not a very lucky one. She completed six war patrols between her commissioning and the end of the war but succeeded in torpedoing and sinking only one merchant ship, the 7000 ton freighter *Empire Lakeland* on March 8th, 1943.

The submarine left its base in Bergen, Norway on February 22, 1945. She had been assigned to a patrol area on the east coast of Canada, there to attack Allied shipping along the convoy routes entering and leaving Nova Scotia's ports. On the night of April 15th/16th the *190* was patrolling the approaches to Halifax Harbour when *HMCS Esquimalt* was conducting her routine sweep of the approaches. Reith believed his vessel had been detected by the minesweeper, and carried out a panicked attack with one torpedo from a stern tube as they motored away from *Esquimalt* without expecting to hit her. The torpedo ran true and blew a massive hole in Esquimalt sending her quickly to the bottom.

After the sinking of the *Esquimalt,* Oberleutnant Reith avoided detection and snuck away from the area of the sinking, but remained on patrol off the Nova Scotia coast. On May 8th, 1945 *U-190's* radio operator received the signal from Admiral Doenitz to surrender. On May 11th, 500 miles east of Cape Race, Newfoundland, Canadian forces caught up with *U-190*. The next day, Oberleutnant Reith officially surrendered to *HMCS Victoriaville* and the u-boat was escorted into Bay Bulls, Newfoundland.

As with *U-889, U-190* was commissioned into the Canadian Navy and became *HMCS U-190*. After testing the u-boat and learning what they could of her workings, the submarine was decommissioned and a plan devised for her disposal.

The RCN conceived "Operation Scuttled" to dispose of the u-boat. The submarine was painted in festive looking red and yellow stripes and towed to a site (believed to be) near where Reith had sunk the *Esquimalt*. At 1100 on the 21st of October, 1947 the Tribal class destroyers *HMCS Nootka, HMCS Haida*, the Algerine class minesweeper *HMCS New Kiskeard,* and twenty aircraft prepared to do battle with the unmanned enemy while a multitude of press and

Above: U-889 *making 17 knots on the surface while undergoing sea trials with the Canadian Navy in 1946.* Photo courtesy of Jeffrey Smith. **Below:** U-889 *&* U-190 *moored alongside three RCN frigates in Halifax Harbour in 1945.*

Author's collection.

radio representatives looked on. High ranking naval officials and their families were aboard the vessels and were looking forward to a long afternoon's entertainment. The actual result was anticlimactic: Instead of the prolonged aerial bombardment of the u-boat followed by the destroyers attacking with guns and hedgehogs, the submarine sank after the first aerial rocket attack. She slipped beneath the surface just nineteen minutes after the exercise had begun.

As with her sister, the *U-190* has yet to be located. Her position is now believed to be quite far removed from the final resting place of the *Esquimalt*. She is believed to be in greater than 300 feet of water.

Before *U-190* was scuttled, her periscope had been removed. Since 1963 it has been displayed in The Crow's Nest officers club in St. John's, Newfoundland. Likewise were other souvenirs taken from the submarine, one of which was her radio room telegraph key which is now owned by a local ham radio operator who still uses it.

S.S. Charlottetown

In 1931 the *S.S. Charlottetown*, official number 154887, was launched and put into service to replace the aging *S.S. Prince Edward Island* on the Tormentine, New Brunswick to Borden, Prince Edward Island ferry run.

The vessel was built in Lauzon, Quebec for Canadian National Railways, and was 342 feet in length by 59 feet in breadth. She was powered by triple-expansion steam engines capable of producing a combined 7000 horsepower, and was designed with an ice-breaking bow to cut her way through the thick ice of the Northumberland Strait in winter.

The ferry could accommodate 800 passengers, 45 vehicles, and 16 rail cars on three sets of tracks. She also boasted a well appointed first-class restaurant for the enjoyment of her passengers.

On June 18th, 1941 the *Charlottetown* was steaming along the south west coast of Nova Scotia, on her way to a slipway in St. John, New Brunswick for her steamship inspection, when she ran aground on Little Hope Island, near Port Mouton. There was no tug available to come to the aid of the stricken ship so the local fishing vessels tried their best. With the *Charlottetown's* own engines turning full astern, the vessel finally came clear of the rocks. The damage to her forward part was so severe however, that she couldn't be kept afloat and sank a short time later in approximately 200 feet of water.

Although many have claimed to know where the bones of the *Charlottetown* lie, a positive location for the ferry is still elusive and to my knowledge the wreck has yet to be visited by divers. Local technical diver Paul Grantham has expended much time and energy in his quest for the lost *Charlottetown*, but has so far come up empty-handed.

Above: S.S. Charlottetown *seen here entering port with passengers on deck.* Photo Paul Grantham collection. **Below:** *The car and rail ferry* Patrick Morris.

Photo courtesy Marine Atlantic.

M.V. Patrick Morris

Another of Canadian National's ill-fated rail and car ferries was the *Patrick Morris*. The *Patrick Morris* was built as hull number 251 by Canadian Vickers Ltd., in Montreal, Quebec. The vessel was built in 1951 to be used as a railroad and car ferry between Florida and Havana, Cuba.

The ship was originally named *New Grand Haven* and was built as a stern-loading car ferry. At 460 feet long and 10,135 tons it was the largest car ferry of it's time. Immediately after delivery, she was converted to a side-loader to be able to carry shipping containers, but three years later was converted back into a stern-loader again, and was then fitted to carry railcars as well as automobiles.

The *New Grand Haven* was taken out of service in 1959 due to the Cuban revolution and lay idle for the next several years. In 1965, the ship was purchased by CNR and renamed *Patrick Morris*. The ferry was put on the North Sydney, Nova Scotia to Port aux Basques, Newfoundland run.

On the evening of April 19th, 1970 the *Patrick Morris* was at her berth in North Sydney when she received a distress call from the fishing vessel *Enterprise*. The *Enterprise* was in danger of going down in heavy seas north of Sydney and Captain Roland Penney of the *Patrick Morris* wasted no time in ordering his ship to sea to aid the stricken fishermen.

While approaching the last position given by the *Enterprise,* a body was spotted floating in the sea by a sharp-eyed lookout. The *Patrick Morris* was ordered astern, and backed up into the huge thirty-foot seas to try and recover the body of the hapless fisherman. While this maneuver was underway, a particularly heavy sea smashed into the ferry's stern door, driving it inwards and allowing the cold North Atlantic waters full access to the inside of the ship. The *Patrick Morris* sank within thirty minutes, sliding stern first beneath the towering seas which had broken through her defenses.

Early on the morning of April 20th three vessels who had heard the mayday from the *Patrick Morris* arrived at the scene of the disaster and picked up forty-seven survivors of the sinking. Joe Slayman, David Reekie, Ron Anderson, and Captain Roland Penney had all gone down with the ship.

A commercial diving company was contracted by Transport Canada to investigate the wreck and try to determine the cause of the stern door's failure. A friend and dive buddy of mine was a young diver's tender on that particular job. The wreck was found to be sitting upright and intact on the bottom with four railcars still sitting on their tracks on deck. The *Patrick Morris* was found to be lying in 270 feet of water with her upper works reaching to within 180 feet of the surface. The wreck site is approximately 8 miles north of Point Aconi, Cape Breton Island and quite near the ferry track on which she traveled for five years.

Although the position of the wreck was located by the survey team and again by a shipwreck researcher in Cape Breton in 2001 with whom I have been in contact, no technical dive expedition has yet been led to the wreck. The commercial divers who inspected the wreck have so far been the only divers to visit her.

Afterword

The wrecks mentioned in the preceding chapters are but a few of those known and dived around the Nova Scotian coast. Many more await the intrepid explorers who'll expend time, energy, and not a little money to discover these relics of our past, moldering in the darkness, forgotten by the public but remembered by those divers who continue to research and search for them.

Wreck diving, and especially deep wreck diving, in Nova Scotia is an equipment intensive, labourious, time-consuming, and mentally taxing proposition and is certainly not for everyone, with cold water, strong currents, and often poor visibility all combining to frustrate the explorer's undertakings. But for those who do brave the underwater world of our rocky, fog-shrouded shores, the rewards are well worth the effort. I sincerely hope that my ramblings here will entice more divers to dip their toes into Nova Scotia's vast underwater history and perhaps also encourage some non-divers to take up the sport which has so consumed my life.

This volume has been a joy to research and produce, but it has also been quite a learning experience for me. A second book is already in the initial stages, which will contain the history and stories of more fascinating shipwrecks of Nova Scotia, and also a few of Newfoundland's favorite wrecks. I hope that you, the reader, have enjoyed this writing enough to join me for more adventures in the future.

Allan P. Henneberry.
Eastern Passage, Nova Scotia
September 19th, 2007

Correspondence to the author may be sent via email to diverberr@hfx.eastlink.ca with Wreck Diving Tales in the subject line.

Acknowledgements

Despite the fact that this is but a modest volume, a great deal of time and research has gone into it. This fact made it necessary to use the resources of not a few institutions and individuals. In thanking everyone below, I hope I have not left anyone out. If I have forgotten anyone I assure you it is my own oversight. Rest assured that your efforts have been greatly appreciated.

My sincerest thanks to those photographers who've graciously allowed the use of their art in this volume. My good friend Michael Grebler, whose photographs are found throughout this book and who also assisted with the editing of the volume and wrote the back cover copy. Thanks also to Paul Grantham for use of his historical and underwater photographs, to Jason Kennedy for the same, as well as assistance with some aspects of research. To Hubert Hall and ShipSearch Marine for historic photographs of hard to find ships. My heartfelt thanks also to Gordon Fader for providing sidescan sonar images of some of the shipwrecks. Thanks to Jeffrey Smith for photographs, as well as finding almost irretrievably lost research materials and providing information of various diving endeavours. Also for photographs I'd like to thank Skipper David Gray, Marine Atlantic, the Government of Canada, and my dear friend the late Paul Crocker.

I'd like to acknowledge the efforts of the Public Archives of Nova Scotia, Library and Archives Canada, Marilyn Gurney at the Maritime Command Museum, the National Maritime Museum in Greenwich, England, The Norsk Sjofartsmuseum (Norwegian Maritime Museum), National Maritime Museum and Vasa Museum of Sweden, the Memorial University of Newfoundland, Robert Ogilvie of the Nova Scotia Museum, the Fisheries Museum of the Atlantic, and the United States Archives and War Records for their kind assistance during my research for this book.

To all those who've kindled and fuelled my love of the underwater world over the years. Stu "Flipper" Beakley—Chief, FDU (A), Ret'd, Terry Preeper, Jim Henneberry, Pete DeGrace, Sam Millett, David Pilot, Gary Gentile, Greg Mossfeldt, and all the divers I've had the pleasure and privilege of exploring the underwater world with. It has been an education, an adventure, and a privilege to dive with and know you all.

To Lyle Craigie-Halkett and Roy Martin for assistance and information pertaining to Risdon Beazley's salvage activities in Canadian waters, and to Alfred LeBlanc for his recollections of the *Halfish*.

To Carrie-Ann Smith and the Pier 21 Museum staff, for support and the chance to once again visit the waters adjacent to Pier 21, and to Gordon Helm of the Halifax Port Authority for granting permission to dive in the waters of Halifax Harbour.

To dive boat captains David Gray (*Ryan & Erin*), Mike Gregg (*Lady Shirleen*), Jim Lawton (*Michael P.*), David Belliveau (*Stormy Weather*), Steven Perry (*Sea Gypsy III*), Bill Flower (*Islander VI*), and Barry & Blair Sullivan (*Badlands*) for getting us to remote shipwreck sites, often in inclement weather. Also, to my brother Ron Henneberry, for the use of his fishing vessels *Lorna J. II.* and *Crabs 'R' Us* for diving operations and shipwreck searches.

To Teri Davies-Charlton, found again after so many years, thanks for pushing me to finish this book when I had gotten lazy and stalled in my writing.

Last, but certainly not least, I'd like to thank my better half Janice for putting up with the obsessions of a rabid wreck diver and offering support and understanding throughout this project.

Thanks to all who've made this book and my diving possible. If I've inadvertently omitted anyone, I apologize for my forgetfulness, but know that your efforts and support have been appreciated.

Selected Websites & Reading

Diving Forums

EC_Divers—Diving forum based in Dartmouth, Nova Scotia.
   http://ca.groups.yahoo.com/group/EC_Divers/?yguid=139945055

NB_Divers—Diving forum based in New Brunswick
   http://ca.groups.yahoo.com/group/nbdivers/?yguid=139945055

Mossman Deep Technical Diving—Expedition reports, DIR Diving, technical diving, discussion group.
   http://www.mossmanscuba.com/

The Decostop—Forum dedicated to technical diving discussions and equipment
   http://www.thedecostop.com/

The Scubastop—Sport diving sister forum to the Decostop
   http://thescubastop.com/

Rebreather World—Rebreather related discussions
   http://www.rebreatherworld.com/forums.php

Dive Shops, Training, & Charter Operations

Deep Down Diving, Professional Dive Training—Technical open-circuit and rebreather training. Based in Halifax, Nova Scotia.
   http://www.deepdowndiving.com

Torpedo Rays Scuba Adventures—Full service dive shop based in Dartmouth, Nova Scotia. Tech training, charters.
   http://www.torpedorays.com/

Skipper Dave's Charters—The premier charter operation in the Halifax area. Day charters or expeditions. Deep-sea fishing. Captain David Gray.
   http://www.skipperdavescharters.com/

Louisbourg Scuba Services—Recreational and Technical dive training and equipment service in Louisbourg, Nova Scotia.
   http://www.louisbourgscuba.com/
Lunenburg Ocean Adventures—Dive charters, *HMCS Saguenay*, shark diving, deep-sea fishing. Captain Bill Flower.
   http://www.lunenburgoceanadventures.com/
Causeway Diver's Supply—Dive shop and training near the Canso Causeway.
   http://medianetcom.com/causeway/
Sand & Sea Dive Shop—Full service dive shop located in Kentville, Nova Scotia.
   http://www.sandandseadiveshop.ca/
Scuba Tech—Full service dive shop in Sydney, Nova Scotia.
   http://www.scubatech.ns.ca/index.html
Dive Centers.net
   http://www.dive-centers.net/dive_shops-canada-43.html

Equipment

Otter Drysuits—High quality suits and accessories for technical and commercial diving. (Contact the author diverberr@hfx.eastlink.ca for Canadian pricing)
   http://www.drysuits.co.uk/
Dive Rite Express—Dive Rite products
   http://www.diveriteexpress.com/
Deep Sea Supply—Innovative scuba equipment
   https://www.deepseasupply.com/
Tech Diving Limited—Great deals on technical dive gear.
   http://www.techdivinglimited.com/
Nuvair—Breathing air compressors and Nitrox solutions.
   http://www.nuvair.com/
Golem Gear—Technical diving equipment.
   http://www.golemgear.com/
Salvo Diving—Technical diving equipment. Specializing in U/W lighting.
   http://www.salvodiving.com/home
Oxycheq—Analyzers, tech equipment, rebreather accessories.
   http://oxycheq.com/Oxycheq/Welcome.html
Maxtec—Analyzers and sensors.
   http://www.o2sensor.com/aboutUs.html
Extreme Exposure—DIR/Technical equipment
   http://extreme-exposure.com/

Underwater Kinetics Canada—Dive lights, Dry boxes.
   http://www.underwaterkineticscanada.com/
AP Diving—Inspiration and Evolution rebreathers.
   http://www.apdiving.com/flash_content/flash_content.html
Innerspace Systems—Megalodon rebreather.
   http://www.customrebreathers.com/
KISS Manufacturing—KISS and Sport KISS rebreathers.
   http://www.kissrebreathers.com/

Research & Shipwreck Information

Library and Archives Canada
   http://www.collectionscanada.ca/index-e.html
Shipwrecks of Nova Scotia
   http://nswrecks.net/
The British Merchant Navy
   http://www.merchant-navy.net/index.html
Ubootwaffe.net
   http://www.ubootwaffe.net/index.html
U-Boat.net
   http://uboat.net/index.html
John H. Marsh Maritime Research Center
   http://www.rapidttp.co.za/museum/jmmrc.html
National Maritime Museum, Greenwich, England
   http://www.nmm.ac.uk/index.php
Dive Nova Scotia
   http://www.divenovascotia.com/photo/
National Maritime Museum—Sweden
   http://www.marinmuseum.se/InEnglish/about.aspx
Norwegian Maritime Museum
   http://www.norsk-sjofartsmuseum.no/pub/index.php?subkat=en&lang=2
Haze Gray & Underway
   http://www.hazegray.org/features/

Periodicals & Books

Wreck Diving Magazine
   http://wreckdivingmag.com/

Diver Magazine
   http://www.divermag.com/online/
Gary Gentile Productions
   http://www.ggentile.com/

Suggested Reading

Time in a Bottle—Bob Chaulk. Pottersfield Press, 2002. ISBN—1-895900-51-4

Shipwrecks of Nova Scotia, Volume 1—Jack Zinck. Lancelot Press, 1975. ISBN—0-88999-042-5. (Out of print)

Shipwrecks of Nova Scotia, Volume 2—Jack Zinck. Lancelot Press, 1977. ISBN—0-88999-076-X. (Out of print)

Wreck Diving Adventures—Gary Gentile. Gary Gentile Productions, 1994. ISBN—1-883056-00-4

Mastering Rebreathers—Jeffrey E. Bozanic. Best Publishing, 2002. ISBN—0-941332-96-9

Wreck Hunter—Terry Dwyer. Pottersfield Press, 2004. ISBN—1-895900-67-0

The Technical Diving Handbook—Gary Gentile. Gary Gentile Productions, 1998. ISBN—1-883056-05-5

Halifax, Warden of the North—Thomas H. Raddall. Nimbus Publishing, 1993. ISBN—1-55109-060-0

The Grey Seas Under—Farley Mowat. Little, Brown & Co., 1958. ISBN—0-316-58637-4

CCR Trimix Simplified—Dr. Mel Clark. 2007. www.silentscuba.com

978-0-595-50050-5
0-595-50050-1

Printed in Great Britain
by Amazon